THE ULTIMATE

NO GRID

I0446635

PROJECTS BIBLE

Master Self-Sufficiency in 2500 Days of
Ingenious Tested Projects.
A Definitive Guide for DIY Enthusiasts to
Overcome Economic Challenges.
Embark on a Self-Reliant Odyssey.

GRANT SIMS

Disclaimer

Medical Disclaimer

Legal Disclaimer

Trademark Disclaimer

TABLE OF CONTENTS

INTRODUCTION

Greetings from the ultimate resource for people looking to live less grid-dependently, with resilience and self-sufficiency. The "No Grid Projects Bible" is more than just a book; it's an informational resource and a survival toolkit for a variety of life circumstances. We cover all the important subjects in this extensive 11-in-1 bundle, giving you the tools you need to face the challenges of today's world with assurance.

1. Surviving the Wild: An Essential Survival Guide

Get ready for anything unexpected. This section gives you the tools you need to survive in the wild, from learning how to build fires to navigating through untamed terrain.

2. Sustainable Living: Reducing Waste and Saving Your Wallet

Learn useful tips for leading a sustainable lifestyle that will save you money and lessen your impact on the environment.

3. Raising Livestock: A Comprehensive Guide to Animal and Poultry Breeding

Set out on an adventure into ethical animal husbandry that will equip you with the skills necessary to raise livestock that are both healthy and productive.

4. Emergency Care Essentials: Medical and First Aid for Survival

Acquire first aid and life-saving skills so that you can confidently handle medical emergencies in situations where professional help is not easily accessible.

5. Solar Power Unplugged: Off-Grid Energy Generation

Leverage solar energy to break free from conventional energy sources. You can set up your own off-grid solar power system with the help of this section.

6. The Modern Homesteader's Handbook: Mastering Self-Reliance Skills

Learn the practical skills needed to become a modern-day homesteader, from food preservation to house upkeep.

7. Wild Edibles: A Forager's Handbook

Discover the abundance of nature by connecting with the abundance present in the natural world with a guide to identifying and using wild edibles.

8. Cultivating Abundance: A Guide to Growing Your Own Vegetable Garden

With these helpful hints and strategies for successful vegetable gardening, you can turn your backyard into a source of fresh, organic produce.

9. Orchard Oasis: Nurturing Your Family's Fruit Trees

Learn the art and science of orchard maintenance to guarantee your family has an abundant crop of tasty and nourishing fruits.

10. Harvesting the Wild: A Guide to Hunting and Fishing

Gain knowledge of hunting and fishing, which will provide you food and a link to the earliest stages of self-sufficiency.

11. Fortify Your Home: A Comprehensive Defense System

Discover how to build a complete defense system and fortify your home to protect the things that are most important to you.

With the "No Grid Projects Bible," set out on the path to self-sufficiency. Regardless of your level of experience, this bundle will help you live a more resilient, sustainable, and satisfying life.

Book 1

Surviving the Wild: An Essential Survival Guide

More than ever, we need to arm ourselves with the knowledge necessary to succeed in a variety of circumstances. Imagine this: being able to survive in the wild for at least 72 hours when things get rough gives you a significant advantage over most people.

Here, we're not discussing a carefree camping excursion. Here's the real deal: you might not even have time to reach for your emergency supplies or your go-to tools in an actual survival situation. But have no fear—hard skills, a resilient spirit, and a strong will to live can help you overcome survival's obstacles.

Being proficient in outdoor survival can significantly improve your chances of success in critical situations. Consider creating a shelter, locating water, gathering food in the outdoors, and starting a fire.

The actual test is about to begin. Put those hard skills into practice with sincerity. Play through the obstacles you might encounter in the real world. It takes more than just physical prowess to solve problems; one must also possess mental fortitude, optimism, and inventive problem-solving abilities. regular practice and advancement? That will give you the confidence you need to face difficult situations head-on.

If you ever find yourself in this kind of world, stay put as we dive right in. What do you know? How do you start taming the wild? Yes, you've got it—master those survival techniques.

Wilderness survival skills

Have you ever wondered what the top priority for survival in the wild is? It isn't the usual suspects like water, food, fire, or even a reliable knife. No, it begins within—with your own mindset. Survival isn't just about what you do; it's also about who you are in difficult situations. Enter the survival psychology.

Let me tell you a story: There was this guy lost in the woods, panicking and running in circles in an attempt to find a way out. What's the catch? He ran straight across a road and into the woods, where he died. Isn't that extreme? But it does make a point: panic can cloud your judgment.

So, what is the main point of this wilderness survival guide? It's not about appearing to be a hero out there. Survival can be difficult—cold, hungry, thirsty, wet, and tired—but it's all about staying calm. What is the top priority? Keeping a level head. Maintain a clear mind and a positive attitude.

The goal is to survive. And what about the first rule? Don't freak out. Maintain a calm and positive attitude while drawing on your training and preparation. Take a moment to evaluate your situation and devise a strategy. What are your resources? What critical tasks must you complete in order to survive on your own in the wilderness? That's where you'll begin.

What happens after you've established a calm frame of mind? What other survival skills will help you out there? These are the essentials—skills that will increase your chances of success in the wilderness. Let's get started.

Build a fire.

Have you ever found yourself in the wilderness in need of a fire but discovering that your trusty firestarter kit isn't quite as reliable as you thought? It can happen to the best of us. But don't worry, knowing how to start a fire without matches or a lighter is a game changer in survival skills. You never know when you'll need to light a match. Plus, how about impressing your camping companions with an old-school fire-making trick? Priceless.

Why is it so important? Simply ask Bud Ahrens. He was leading a dog sledding trip in Minnesota when a teammate fell through the ice into freezing water. They had warmth in 20 minutes thanks to quick thinking and knowing how to build a fire, potentially saving a life.

So, why bother learning how to start a fire if you don't have the necessary tools? Because it's not just about cooking or staying warm when you're off the beaten path; it's about survival. This section? It's your go-to resource for the top six wildfire starting methods. Spoiler alert: fire requires fuel, oxygen, and a spark. Let's get started.

1. The Hand Drill

Starting a fire with sticks? Here's the lowdown on the hand drill method—it's a bit primitive but can be a survival game-changer.

Tinder Nest: Begin with a nest made of things that catch fire easily—dry leaves, grass, or bark work like a charm.

Notch Creation: Cut a slanted notch on your fireboard and make a little depression near it.

Bark Placement: Slip bark under the v-shaped cut to catch the ember produced by the friction between the spindle and fireboard.

Spinning Action: Put the spindle into the groove on your fireboard (about a two-foot spindle works best). Apply pressure and start rolling the spindle quickly in your hands down the groove until you see an ember on the fireboard.

Fire Ignition: Once you spot a glowing ember, tap the fireboard to transfer the ember onto a piece of bark. Move the bark to your tinder nest and gently blow to get that fire going.

2. Fire Plow

Minimal tools needed for this one—just a softer wood for the plow board and a harder wood for the plow itself.

Setting Up: Create a fireplace and carve a groove about one inch wide and six to eight inches long down the board.

Carving the Plow: Take a piece of hardwood about one foot long and carve it to a point at the end.

Rubbing Time: Place the pointed end of the plow on the groove of the fireboard. Rub the tip back and forth in the groove, creating small dust pits.

Board Elevation: Lift the top of the board and rest it on your knee to collect the dust at the bottom.

Intense Rubbing: Once a small pile of dust has gathered, rub the plow vigorously in the groove until the dust starts smoldering.

Flame Transfer: When the dust is smoldering, transfer it to your tinder and gently blow to kickstart your fire.

3. The Bow Drill

The bow drill is a winner when it comes to making fire without matches because it is contact-based and effective.

Materials Required:

A fireboard is a flat, softwood board that is at least a foot long and six inches wide.

Socket: A flat rock with a depression in which to rest the drill.

A sturdy hardwood stick about a foot long for drilling.

A bow is a two-foot-long, sturdy, flexible green stick.

Cord: Paracord or boot laces both work well.

Fire Procedures:

Make Your Own Bow: Tie the cord in a half-moon shape around the bow.

Install the Fireboard: Underneath, make a shallow depression and a V-shaped notch.

Make a Drill String: Circumvent the drill with the bowstring and apply pressure with the socket.

Begin sawing: Move the drill in a sawing motion, pivoting quickly. Continue until you have created an ember.

To start a fire, place an ember in a nest and gently blow on it.

4. Steel and stone

No lighters or matches? Not a problem! How to Start a Fire with a Flint and Steel.

- Create a Tinder Nest: Create a nest to catch the flint and steel spark.
- Gather your starters: With one hand, hold the steel and the striker.
- Strike: Hold the steel steady and smoothly move the striker down its length.

- Start a Fire: Insert the spark into the nest and gently blow to start the fire.

5. Batteries and Steel Wool

Here's an easy way to start a fire with things you probably have in your backpack—batteries and steel wool.

Steps:

Stretch the Wool: Make the steel wool about 6 inches long and at least an inch wide.

Rub with Battery: Take any battery (a 9-volt works best) and rub it on the steel wool. The wool will start to glow and catch fire.

Set the Fire: Place the burning wool on your tinder nest and gently blow to get a bigger fire.

6. Traditional Lenses

Who needs matches when you've got sunlight and a lens? Here's how to start a fire with a magnifying glass or binocular lens.

Steps:

Prepare Tinder: Get your tinder ready; it's about to catch fire.

Hold the Lens: Position the magnifying glass between the sun and the tinder. Look for the bright dot that shows up. Tilt the lens so the dot is over the tinder and about a quarter-inch in diameter.

Focus the Dot: Concentrate on the dot for 30 seconds to a minute. Patience is key! Once the tinder starts smoking, gently blow to get the flame going.

Remember: Learning to light a campfire without a lighter or matches takes practice. Don't wait for a life-or-death situation to try these out. Keep practicing and stay prepared!

Survival shelter

Millions of TV viewers have been captivated by the idea of being a "survivor." But a survivor is not just a fantasy from television. When life away from home doesn't go precisely as expected, a survivor is someone who is ready to live—and live as healthfully as possible.

Knowing what to prepare for is the first step towards being prepared to survive in the outdoors. You can survive for weeks without food and for days without water. Not always from malnutrition or thirst, but more frequently from losing their body heat, those who fail to survive in the outdoors die. To keep your body heat contained close to your body, where it belongs, and to fend off wind, rain, and snow, you must be able to construct a shelter.

The following are essentials for seeking refuge in the wild:

DRESS WISELY

Wearing a "shelter" is your first line of defense against the weather. You are prepared for anything if you carry a shell made of windproof and waterproof material and wear layers of wool or synthetic material. Your body heat will be retained rather than released into the environment.

THE IDEAL PLACE

It's crucial to pick the ideal location for your survival shelter. It ought to be in the driest area you can discover. Nothing removes body heat as quickly than moisture. Make a shelter on high ground if the weather isn't too chilly. In addition to helping to keep pests away, breezes will make you more visible to search parties passing by. Select a location that is protected by trees if there is a strong wind. However, avoid building near the base of steep valleys or ravines where nighttime cold air condenses.

SLEEP BED

Until you have a bed to lie in, your shelter is not complete. Crisp leaves are ideal. Make sure your bed is at least eight inches thick and slightly larger than the area your body occupies. You may curl up with it and be prepared for that last-minute night out.

Types of shelter and how to build them.

THE COCOON

Make a mound two or three feet high and longer than you are tall if you can quickly gather dry material (leaves, pine needles, bark) from the forest floor before it becomes too dark. You are in a natural sleeping sack that prevents heat loss when you burrow into the pile.

THE FALLEN TREE

A fallen tree with enough space below it for you to crawl inside is the most basic kind of shelter. To create a wall, lean branches against the tree's windward side, which faces the direction of the wind. Build a wall that is thick enough to block off wind. You can stay warm if you can create a fire on the exposed side of your shelter.

THE LEAN-TO

You can construct a basic lean-to if you come upon a fallen tree with enough space underneath it, a rock, or a little overhang. To begin, build a wall by laying fallen branches against an item, such the upper edge of an overhang. To assist protect against rain, lean the limbs at an angle. Use everything the forest has to provide to cover the drooping branches, such as leaves, boughs, pine needles, bark, etc. You can enter your shelter by crawling underneath once you have constructed a substantial wall. Don't forget to build your shelter no larger than what is necessary to accommodate you and any more people. Maintaining warmth becomes more difficult in larger spaces.

Another way to construct a lean-to is to lay one end of a long stick over a low tree limb and support the other end of the stick with two other sticks. Use your belt or boot laces to knot the ends of the sticks together. Press more sticks up against the horizontal one. Then, until you create a wall, stack leaves and other forest detritus against the leaning sticks. Once again, your "room" will get a lot hotter with a fire on the open side of the lean-to.

THE A-FRAME

An A-frame shelter can be constructed in the event that a lean-to is not feasible. Two sticks, each four or five feet long, and one stick, ten or twelve feet long, are required. Raise the two shorter sticks so that they resemble the letter A. At the top of the A, prop up the longer stick. Where the three sticks converge, tie a knot. With one end of the three sticks folded on the ground, they will resemble an A-frame tent. Once you have an insulated shelter open at the top, continue stacking sticks against the longer stick and adding forest material to them.

A TARP

Tie a line between two trees using a tarp, plastic sheet, or Space Blanket that you have with you, along with some rope or cord. Make sure you tie it low enough so you can sleep beneath it. Over the line, extend the tarp. To keep the tarp in place with its edges close to the ground, lay heavy boulders or logs on its ends. Tie the line off higher on the trees if it's snowing. Better at shedding snow are steeper walls. You now own a tent for emergencies.

Inappropriate locations for a shelter in the wild.

- Anywhere there is wet ground.
- In areas with exposed ridges and mountaintops where the wind is brisk.
- At the base of little valleys where the cold gathers at night.

- Washes or ravines where water flows after a downpour.

Find clean water source.

No shelter. Absent food. Not a fire.

These won't kill you, not fast anyhow, unless circumstances are really dire. However, all of these pale in comparison to the scarcity of water, which has far greater relevance. You probably won't survive more than three days without water, but you can survive for more than three weeks without food. While some individuals have managed to last up to ten days without water, most people's capacity to operate drastically diminishes after the third day.

Every day, our bodies require two to three liters of water. Add in the elements—heat, cold, stress, exercise, or diarrhea—and you'll require a lot more. You must understand where to get water and how to keep your body from dehydrating.

Drink all that you can.

If they're fortunate enough to locate water, one concern that people have with it is whether or not they should drink it at all for fear of becoming sick. Keep in mind that the effects of drinking unclean water will not cause you to die any sooner than dehydration. Actually, drinking untreated water won't kill you unless there are really unusual situations. If you survive, you can use potent medications to treat any illness or parasite.

Having saying that, you should never take water lightly. Drinking tainted water might cause you to pass out from agony and diarrhea in a matter of hours, making your survival struggle much more difficult. Treating the water and assuming that it is tainted is your best option. You ought to filter your water if it is something you can do. However, if you have to choose between dying of thirst or drinking untreated water, then drink.

Finding Water

When searching for water, make an effort to preserve what you already have and look for a replacement source as soon as you can. The principal water sources that flow are the best. Rivers, streams, and creeks are a few of these. After that, you start to approach more still water features, such as lakes and ponds.

Once you've located a water source, search upstream or along the shoreline for pollutants like dead animals. You are getting closer to the cleanest water—one that hasn't been contaminated by contaminants or decomposing matter—the further up the water table you are.

Remember that even the purest-appearing mountain streams may include invisible pollutants upstream; this is just another reason to boil or filter your water whenever possible.

Your best choice is to research the terrain of your surrounds in order to discover a primary supply of water. Generally speaking, walking downhill is a smart move. Water is easily found at valley bottoms because it flows downhill freely.

Natural Indicators

Keep an eye out for plant changes, since these signify the presence of water. You have a decent chance of finding water if you notice an area where the vegetation is thicker or darker than the surrounding region, even if you have to dig for it.

Another little tactic I frequently employ in survival circumstances is to notice minute variations in the hue of the sky. The area of the sky immediately over a body of water will usually appear bluer than the surrounding air. Low-lying clouds and fog also tend to gather over a body of water early in the morning. A body of water not only reflects the sky differently than a dense forest, but the fog is also created by the water's moisture content and temperature differential.

Help From Wildlife

Though they may guide you to water, animal tracks might sometimes take you to nothing. If there are many game paths, they may develop a structure like a network of veins. The tip of the "V," formed by the joining of the pieces, may point towards the water.

Remember that the majority of wild animals relieve themselves in the same areas where they drink. Move at least a few hundred yards away from the location where the game route meets the water once you've found it, ideally upstream.

Birds also prefer to gather around bodies of water; early morning or late afternoon bird flights might reveal the direction of bodies of water. Water is always nearby for grain-eating birds, so if they are flying low and straight, they may be approaching it. Insect swarms may also be an indication that water is nearby. There may be water in a tree hole if you observe bees or ants entering it. The water may be sucked out using plastic tubing, or it can be absorbed by stuffing a towel within the opening.

Wintry Water

You have access to an excellent supply of water if you happen to be in an area or at a season where there is ice, slush, or snow, especially if you know how to make fire. A lot of survival teachers would advise against eating snow, mostly because it can lower your body temperature and burn up valuable energy while warming up. Although this is true, I think otherwise because of how essential water is to life.

Eating snow can help you retain body warmth rather than dangerously cool down if it's early in the morning and you're working hard to ensure other parts of your survival. You also need that valuable liquid.

Later in the day, when you're exhausted and the weather is starting to chill down, you have to be cautious while consuming snow and ice. This is true not just in the middle of winter, but also in the spring when you consume snow. Your body's defenses are weak at this point, so you run the risk of doing more damage than good.

The best scenario, of course, is to heat the snow and ice before drinking it. In the event that you are unable to start a fire, place a filled water bottle with snow below your clothes while working during the day or under your sleeping bag while you are asleep at night. The initial portion takes a long to melt, but when it does, the remaining portion melts significantly faster. It would be nice to wake up to melted water that is ready to drink, if I can do this without becoming too cold.

Food source in the wild.

Food is man's most pressing need, after water. When one considers almost any fictitious survival scenario, food comes to mind right away. Even water, which is more vital to sustaining bodily processes, will nearly always come after food in our minds, unless we are in a dry region. It is imperative for the survivor to have in mind that the hierarchy of needs for water, food, and shelter is contingent upon the survivor's assessment of the current circumstances. This estimate needs to be precise as well as timely.

Animals for Food

Due of their abundance, focus your efforts on the smaller animals until you have the opportunity to take huge game. It's also simpler to prepare the smaller animal species. You must be ignorant about all the animal species that are fit for human consumption. Few of them are toxic, so that's one less thing to keep in mind. Understanding the behaviors and behavioural characteristics of different animal types is crucial. Animals with fairly stable feeding regions, those that inhabit a certain range and occupy a den or nest, those that make ideal trapping candidates, and those with trails connecting various places are a few examples. Larger herding animals that cover large regions, like caribou or elk, are more challenging to catch. You also need to know what foods a certain species like to eat.

With very few exceptions, everything that crawls, swims, walks, or flies may be eaten. Overcoming your innate resistance to a specific food source is the first challenge. People have historically eaten everything and everything to survive during times of hunger. A person puts his own existence at risk when they disregard a generally healthful food source because of a prejudice or because they don't think it tastes good. To be healthy, a survivor has to consume whatever is available, even if it might not be easy at first.

Insects

Insects are the most prevalent living form on Earth and are quite simple to catch. Compared to 20% for beef, 65–80% of protein comes from insects. Because of this, insects are a significant source of food, albeit not very tasty. All adult stinging or biting insects should be avoided, as should hairy or vividly colored insects, caterpillars, and insects with strong smells. Steer clear of spiders as well as common disease-carrying insects like flies, ticks, and mosquitoes.

An abundance of insects may be found on decaying logs that are left on the ground, such as termites, ants, beetles, and grubs, which are the larvae of beetles. Insect nests on or in the ground should not be disregarded. Fields and other grassy places are ideal places to look for insects since they are more visible there. The insects choose suitable places to build their nests on stones, planks, or other things that are left on the ground. Visit these websites. Larvae of insects are also edible. Parasites are found in insects with a hard exterior shell, such as grasshoppers and beetles. Make them before you eat. Eliminate any wings and stubby legs as well. Most insects are edible when uncooked. Different species have different tastes. While certain kinds of ants store honey in their bodies to give them a pleasant flavor, wood grubs are tasteless. An assortment of insects can be ground into a paste. They can be combined with food plants. You may enhance their flavor by cooking them.

Worms

Annelida worms are a great source of protein. Look for them on the ground following a rainstorm or dig for them in moist humus soil. Once you have them captured, let them sit for a little while in fresh, drinkable water. You can consume the worms uncooked once they have finished washing themselves out or purging spontaneously.

Crustaceans

The size of freshwater shrimp varies from 0.25 to 2.5 millimeters. They may establish rather sizable colonies on pond and lake mud bottoms or in mats of floating algae.

Similar to sea lobsters and crabs are crayfish. Their robust exoskeleton and five pairs of legs—the front pair with large pincers—allow you to identify them. Although crayfish are most active at night, you may find them during the day by searching beneath and around stream stones. Another way to locate them is to search the soft mud in the area around their nests' chimney-like breathing apertures. By connecting internal organs or pieces of offal to a thread, you may catch crayfish. Before the crayfish has a chance to release the bait, drag it to shore as soon as it grips it.

Seafood such as lobsters, crabs, and shrimp may be found from the edge of the surf to ten meters down. At night, shrimp may congregate near a light so that a net may be used to catch them. Using a baited hook or trap, you can catch crabs and lobsters. When you set bait near the edge of the surf, crabs will come to it so you may net or trap them. Since they are nocturnal animals, it is preferable to catch them at night.

Mollusks

Octopuses and freshwater and saltwater shellfish, including bivalves, clams, mussels, barnacles, periwinkles, chitons, and sea urchins, are included in this class. In any water condition, bivalves that resemble our freshwater mussels and terrestrial and aquatic snails may be found globally.

In the rivers, streams, and lakes of the northern coniferous woods, river snails, sometimes known as freshwater periwinkles, are common. These snails can have spherical or pencil-point shapes.

Look for mollusks in the shallows of freshwater, particularly in areas with muddy or sandy bottoms. Seek for their black, elliptical, open valve slits or the thin paths they make in the dirt.

Look in the moist sand and tidal pools near the sea. Shellfish typically cling to rocks that form reefs in deeper water or that run along beaches. From the low water mark upward, seaweed and rocks are home to limpets and snails. Chitons, which are large snails, cling tenaciously to rocks above the sea line. Typically, mussels congregate in large groups at the bottom of rocks, on logs, or in rock pools.

Fish

Fish is a wonderful source of fat and protein. They provide the survivor with a number of clear benefits. They may be obtained silently, and they are often more common than animal fauna. You have to understand fish behaviors if you want to capture them successfully. Fish, for example, typically eat extensively before to a storm. When the water is dirty and bloated after a storm, fish are unlikely to eat. Fish are drawn to light at night. Fish will rest where there is an eddy, such next to rocks, during strong currents. Additionally, deep pools, overhanging vegetation, submerged foliage, logs, and other items that provide cover are among the places where fish will congregate.

Poisonous freshwater fish do not exist. On its dorsal fins and barbels, the catfish species does, however, have pointed, needle-like projections. These can cause excruciating puncture wounds that get infected fast. To eradicate parasites, cook any freshwater fish. As a precaution, roast saltwater fish that has been captured near a reef or near a freshwater source. Because of the salty environment, marine life found further out in the sea is free of parasites. These can be consumed uncooked.

The flesh of several saltwater fish species is toxic. The poisoning is seasonal in some species and chronic in others. Porcupine fish, triggerfish, cowfish, thorn fish, oilfish, red snapper, jack, and puffer are a few types of saltwater fish that are venomous. Although not hazardous in and of itself, uncooked barracuda can spread ciguatera, or fish poisoning.

Amphibians

It's easy to find salamanders and frogs near freshwater bodies. Rarely do frogs leave the protection of the water's edge. They dive into the water and hide themselves in the mud and debris at the first sight of danger. There aren't many toxic frog species. Steer clear of frogs with vivid colors or those with a noticeable "X" mark on their back. Toads and frogs are not the same thing. Toads typically inhabit drier settings. A toxic material is secreted by the skin of certain toad species as a protection mechanism against predators. Therefore, avoid handling or eating toads to prevent poisoning.

Salamanders live at night. Using a light, nighttime is the ideal time to capture them. Their length can vary greatly, from a few centimeters to well over sixty. Look for salamanders in the water along the mud and rock banks.

Reptiles

Reptiles are rather easy to capture and a rich source of protein. Although they are best cooked, you may consume them raw in an emergency. Although parasites can be spread through their raw flesh, reptiles can not carry the blood illnesses that affect warm-blooded animals.

One frequent type of turtle you shouldn't eat is the box turtle. It consumes deadly mushrooms for sustenance, and its body may accumulate a very dangerous toxin. This poison is not eliminated by cooking. Steer clear of the Atlantic Ocean-dwelling hawksbill turtle due of its toxic thoracic gland. Large sea turtles, crocodiles, alligators, and poisonous snakes provide clear risks to the survivor.

Birds

Birds of any species can be eaten, however the flavors will differ greatly. Fish-eating birds can have their flavor improved by skinning them. Like any wild animal, your chances of successfully catching birds are greatly increased if you are aware of their typical behaviors. Pigeons and a few other species can be manually removed from their roost at night. Certain species will not leave their nests during the breeding season, no matter how close they are to it. It is simpler to catch birds if you know where and when they nest. Flyways that lead from the roost to a feeding spot, water, and other locations are often routine for birds. These flyways should be easily identified by careful observation, which will also suggest ideal locations for netting birds. The most attractive places for snaring or trapping are waterholes and roosting spots.

Eggs from nesting birds are another source of food. Mark the eggs you choose to keep and remove all except two or three from the clutch. In order to complete the clutch, the bird will keep laying eggs. Remove the new eggs one by one, leaving the ones you highlighted.

Making a weapon in the wild.

You will most certainly need to shift your mindset from the 21st century to the prehistoric era when it comes to crafting weapons in the wild. Think primitive. For thousands of years, people used handcrafted tools and weapons manufactured from natural materials to both defend and feed themselves. And you'll have to do the same thing if you find yourself stranded in the middle of nowhere with nothing but the shirt on your back.

Fortunately, assembling a personal arsenal really calls for a small number of items, most of which are readily available wherever you go. Look for sturdy materials such as wood, stones, and animal bones to use as the instrument's hardware.

A knife is the one crucial item you should concentrate on first in order to build up your armory. A knife is a very useful tool in survival scenarios due of its versatility.

Stone holds up the longest for knife blades, while wood and bone also work well.

To craft the blade, three necessary instruments are required:

- Core rock. The actual blade will be a large, comparatively flat stone. Let it to expedite the process of shaping.
- A big, polished rock called a "hammer stone" is used to remove the blade's form from the center.
- Pressure flaker. a sharp-pointed antler or rock used to sharpen the edges of blades.

If you come upon a huge animal bone, break it up with a rock and make a sword out of one of the fragments. A bone knife is useful for punctures, although it is not meant to be used frequently. Similar to steel blades, wooden ones may be made stronger by utilizing only wood with straight grains and drying it over a low flame.

Now that you have a blade, you will get cuts if you handle it in your bare hands. How may a knife's handle be attached?

To finish your knife and other weapons, you will need to produce your own binding material because duct tape is not something you will find in the woods.

The strongest natural rope found in the woods is found in animal tendons. Once an animal has been skinned, take out the sinew and let it air dry. After it has dried, stomp it to separate the tendons into separate cords. Twist two sets of strands in a clockwise way to strengthen them. Next, take the top two bundles and wrap them around one another in a counterclockwise direction. Additional materials that can be used as lashings include strings of plant fiber from tree inner bark and rawhide.

Now that you have rope at your disposal, haft your knife by binding a wooden or bone hilt. Make a cut so one end of the blade will fit in the depression on the end of your hilt. Using your rope, firmly fasten the blade and hilt together.

Additionally, you may put together a weighted club to bash and hammer slow-moving creatures like rats, possums, and porcupines. provide a V-shaped cut in the top of your club with your knife, or use a solid stick with a forked end to provide a room for a spherical rock. Tie it down with your own rope.

Basic First Aid

If you become ill or are hurt while you're in the wilderness, it may be challenging for rescue personnel to get to you. Knowing how to provide first aid and being able to do so might save your life or the life of someone you love.

Don't panic first. According to Tod Schimelpfenig, curriculum director of the Wilderness Medicine Institute at the National Outdoor Leadership School, "you can improvise practically anything in your first-aid box."

Wound Cleaners

Simple water is a very effective wound cleaner, according to research. Use a bladder or zip-top bag to dispense one liter or more of the purest water available to irrigate the injury; if soap is available, apply it to the surrounding skin but not the wound itself, and rinse with water afterward. If not, bandage the area until you can get antiseptic again.

Bandage

Take the cleanest cloth that is available, rip off a section, fold or crumple it, and lay it over the incision. Put in some force. If extra bandages are required, put them on without taking off the previous ones. Clean the area and apply a new bandage after the bleeding has stopped. Use straps or cloth strips to secure the fabric in place. Use ¼-inch duct tape strips to seal gaping (nonvenomous) wounds as closely as possible to the original skin location.

Splints

Creativity is the key. A stove screen, long bundles of grass (line them lengthwise down the limb), hiking poles, twigs, pack supports, or your sleeping mat are a few excellent options. Use fabric strips, straps, or vines to hold the splint in place. To immobilize the joints above and below the fracture, make sure the splint is cushioned yet firm. For example, if you break your shinbone, you should immobilize your knee and ankle. In order to evaluate circulation, the splint should not impede blood flow and should leave room for fingers and toes.

Nature's antibiotic

For millennia, people have used usnea, or old man's beard, as an antiseptic. All around the world, these hair-like, greenish tufts grow on tree branches. Unfold the sheath covering the main stem; the usnea has a white cord at its core. Put a bunch on the incision.

Signal nearby search and rescue teams.

If you don't have access to any kind of modern technology, the most popular and efficient way to signal for assistance is with a signal fire. A well-constructed signal fire will draw interest from people for kilometers around. It also has the added benefit of telling an aerial rescuer (a helicopter, for example) about the wind conditions where you are. But in order for your signal to be as successful as possible, you'll want to make sure you get these distinctions perfect. A good signal fire is different from your typical campfire or cooking fire in a few key ways.

You will need to assess your resources first. Keep your signal fire burned as long as possible and as often as you can if you are in an area where dry wood is plentiful. On the other hand, you would be far better off building a pyre and waiting for the right opportunity, like seeing a search and rescue plane, if you are in a place where fuel is scarce. If at all feasible, position the pyre(s) in a wide, open space on high ground where people may plainly see them. Think about constructing three of these pyres and arranging them in a triangle, around 100 meters apart. Both three and the triangle arrangement are recognized as global indicators of distress.

When building these pyres, you need take care to ensure that the wood remains dry and ready for lighting. Moreover, you should confirm that they can be ignited right away, if at all feasible. Create an elevated platform for your fire to do this. Three long, straight boughs should be bent into a teepee shape, and the top should be secured with wire, string, or vine. Then tie cross-thatched branches to the three supports to make a platform halfway down the branches. Now that you have this support structure in place, you can start adding gasoline.

For the first layer, you want to have nice, dry tinder. An abandoned bird's nest, paper, wood shavings, or dried grass can work well as tinder, if you can locate one. Apply a thick layer of tinder first, and then top with little wood kindling in the shape of broken branches. The fuel should get bigger as you add layers, much like in a conventional fire. You may cover your primary fuel wood with a layer of peat moss, damp leaves, or other decomposing plant material to create a fire that burns more slowly while producing a lot of smoke.

The last layer should be made up of dense, green brush or foliage as your signal fire needs to emit a lot of smoke. When put to a fire, green, live brush almost instantly produces a dense white smoke. In order to create a dense plume of black smoke that is more visible on a cloudy day, you can throw tire rubber and crankcase oil on top of your stranded car.

Creating smoke is less essential at night; what you want are big, obvious flames.

Finally, make use of everything available to you. A survivor has to move fast and pack light, which means they have to carry multifunctional objects that are worth their weight in utility and gather food and water while they're at it. For example, a stranded traveler may not have a constructed bug-out bag or first aid kit from which to gather supplies or equipment needed for a comfortable survival. Situations such as this call for creativity and strong problem-solving abilities. A plastic bag, some paracord, and some duct tape, for instance, may be the main components of a shelter that protects you from the rain and wind.

Emergency preparedness

Should the last several years impart any knowledge, it is that unforeseen circumstances might arise at any moment. Aside from global pandemics, there are an incredible number of natural catastrophes every day that are caused by weather, climate, or water hazards. In the last 50 years, this number has climbed by a factor of five. And in the upcoming years, climate experts only anticipate this number to rise.

Being ready is the greatest approach to safeguard your family's health and safety, even if it may be unsettling to consider. While advice may change depending on the specific emergency scenario, there are certain fundamental steps you and your family should take to be ready for any natural catastrophe. The following are the essential techniques for emergency readiness:

1. Together with your family, devise a strategy.

Creating a detailed strategy is the greatest approach to be ready for unforeseen events. When a natural catastrophe happens, you could not be with your friends, family, or support system, so it's a good idea to get together and plan how you'll get in touch and reunite if you're separated during an emergency. Although each group will have a distinct agenda, setting up a comfortable and convenient meeting location is a good place to start. Other things to think about while creating a plan for your family or friends are:

How are all of you going to be alerted and warned about emergencies?

What is your plan of escape?

Which members of your group—children, senior family members, people with special dietary needs, etc.—have what unique demands?

A list of crucial phone numbers and email addresses, such as those for your family, the local police and fire agencies, or medical providers, could also be included. This worksheet from the Federal Emergency Management Agency (FEMA) is a wonderful place to start if you're not sure how to go about gathering all the pertinent information you'll need to talk about it as a group and come to a consensus. Regardless of how your plan is designed, remember to print copies for every member of the family and to update it as needed. Practice meeting at the agreed-upon meeting place and role-play the different assigned tasks for every member of the family.

2. Stock up on essential basics

You might not be able to leave your house or go to a supermarket or pharmacy for several days if a natural disaster strikes. This implies that, just in case, having a backup supply of food, water, and other necessities is crucial. Ensure that you always keep spares of these on hand in your home:

- For many days, each individual should drink one gallon of water each day for hygiene and drinking.
- A store of non-perishable food that will last for many days (such as protein bars, canned foods, dry cereal, dried fruit, etc.).
- Opening can
- prescription drugs* and eyewear
- Water and food for pets, if needed
- Traveler's checks or cash
- Kid first aid
- An extra battery or charger for your phone
- luminous
- Extinguisher for fire
- A NOAA radio and a hand crank or battery-operated radio climate radio with alarm tone

- Matches in a watertight box
- Personal hygiene products and feminine supplies
- Vital records in a transportable, waterproof container, such as passports and insurance policies
- Whistle to request assistance
- nearby maps
- mask of dust
- If needed, use pliers or a non-sparking wrench to switch off the utilities in your house.
- For kids, books, puzzles, or games if needed
- Paper and a pencil

3. Maintain complete emergency supplies in any location you often visit.

You never know where you might be or when a natural calamity will strike. Even if you can't always carry everything on your list, it's advisable to have the necessities in both your house and car in addition to your place of employment. A few non-perishable foods, water, and other essentials like prescription drugs and cozy walking shoes should be in your work kit in case you have to flee. You might also wish to store the following goods in the trunk of your car, in addition to a few items you've chosen from the above list:

Jumper wires

Cell phone charger for car

Reflective triangles or flares

Chart

An ice scraper

Coverlet

Sand or cat litter (for improved tire traction)

Additionally, it's a good idea to get your automobile serviced on a regular basis by a technician who can check the antifreeze levels, brakes, oil, battery and ignition systems, and more to make sure your car is ready for any emergency. A full gas tank can also assist prevent the fuel line from freezing, so always keep it full in case you need to leave or experience a power outage.

4. Think about your pets.

Although it may seem that your pet will be the first item you grab in an emergency, animals can quickly become disoriented in the turmoil caused by natural catastrophes. Since pets are vital family members, having a plan for them is equally crucial. Create an emergency supply kit with essential drugs, food, and water, just as you would for your family. As required, don't forget to add a carrier, food bowl, leash, harness, plastic garbage bags, or litter. Additionally, you ought to compile a list of a few evacuation locations for your pet. In case of an emergency, be aware of a few pet-friendly places you may travel to, as many public shelters and hotels do not allow animals. If you are unable to take your pet with you, you may also want to consider reaching out to neighbors, friends, or relatives to see if they would be willing to look after your pet. Lastly, get your pet microchipped or tagged in case they ever lose their way or become separated from you.

By following these actions, you may potentially prevent a natural disaster and feel better prepared and in control in an emergency. Visit ready.gov, the official website of the U.S. Department of Homeland Security, if you require more information or are seeking for other methods to maintain your emotional stability.

Navigation

Making do with what we have on hand is frequently necessary for survival. We should be able to navigate across any terrain, day or night, without the need for pricey instruments, even if it would be good to have an electronic GPS device with an endless supply of batteries available.

After all, we could need to flee rapidly with just a little bag of goods in the case of a catastrophic disaster. You may easily waste hours or days aimlessly wandering about in circles if you don't have a reliable method of navigation. The fundamentals of wilderness navigation methods are covered in this handbook.

Using a Compass for Wilderness Navigation

A compass is an incredibly useful and trustworthy navigational tool in the woods. One or more of these ought to be in every person's bug out bag.

There's simply no reason why you and your family shouldn't have a couple of these positioned strategically among your survival supplies, given their compact size and low cost.

The capacity to appropriately utilize a compass is just as vital as possessing one. With any compass, it's important to keep in mind that the red portion of the needle is always pointing northward.

Simply spin the compass housing until the desired direction is completely lined with the "Direction Traveling" line, then rotate your body till the red portion of the compass needle points to the N on the housing in order to go in a direction other than North.

Even though it seems easy, you'd be surprised at how many individuals make this mistake and end up going in the wrong way. To avoid getting erroneous readings, also be sure to take fresh readings from your compass frequently and to keep it away from any metal items you may be carrying or supplies.

Alternative Wilderness Navigation Techniques

While it's ideal to always have a compass on hand for navigating, this isn't always feasible. It may have been lost during your hasty departure to safety, or it might have inadvertently been left behind.

In any case, in the unlikely event that you find yourself in a survival situation, it is critical that you know how to navigate without a compass.

Fortunately, there are many of ways to travel with only rudimentary navigational skills both during the day and at night.

Finding Your Direction with the Sun.

The sun always rises in the East and sets in the West, notwithstanding occasional seasonal fluctuations. When the sun reaches its greatest position in the sky, it is straight south in the northern hemisphere. When there are no discernible shadows produced by objects, the sun is at its highest position.

The sun may be used as a navigational aid in two main ways. While the second employs a watch, the first depends on shadows.

Shadow Tip Method.

Use this way to obtain your bearings: locate a straight stick that is about three feet in length. Choose a level area where the stick may create a distinct shadow that is comparatively free of bush.

Stick the stick into the ground, then use a stone, twig, or just a mark in the ground to indicate where the shadow's tip is. No matter where you are in the globe, this emblem is representative of the West.

Ten to fifteen minutes should pass before the shadow's tip moves somewhat. At this moment, also mark the shadow's tip.

Draw a straight line that roughly runs east-west between the two markings.

You may face north by positioning yourself such that the first mark is on your left and the second mark is on your right.

The shadow tip method is a simple and convenient approach to obtain an estimated orientation, even if it is not 100% correct everywhere in the world.

See Method.

The sun may be used to identify your orientation if you know exactly what time it is. The simplest way to employ this approach is to use a hands-on analog watch.

If all you have is a digital watch, you may do the same thing by drawing a circle on paper that has the proper time on it.

You must use the actual local time (not daylight savings time) while using this approach. For correct results, you will need to make a little adjustment if your watch is set to daylight savings time.

Point the hour hand toward the sun while holding the watch horizontally.

Calculate the angle (split in half) that exists between the watch's hour hand and the 12 o'clock position. Your North-South line is this.

The sun rises in the East, sets in the West, and is straight south at midday, so keep that in mind if you're not sure which end of the line is north.

To adjust for the time difference if your watch is set to daylight savings time, divide the angle between the hour hand and the 1 o'clock position.

Using the Moon

During the night, you may also utilize the moon to acquire an approximate East-West reference. It isn't as exact as some of the other techniques, but if you have no other alternatives, it will work until daylight provides more accurate possibilities.

In general, the west will be lighted if the moon rises before the sun has set. The East is lighted if the moon rises after midnight.

Animal and insect bites.

Attending first aid and CPR courses provided by the Red Cross or another respectable organization is the best way to be ready for any diseases or injuries that may arise while you are traveling. A component of your preparation should be to invest in first aid and CPR training if you want to spend any time in the outdoors. You should never think of this kind of instruction as a choice that you may accept or decline.

This chapter's content will provide you with a basic introduction and sufficient knowledge to get you started; it is not intended to take the place of medical skills that you would learn from a professional in a first aid or CPR training.

Treating Insect and Spider Bites

You should always keep antihistamines in your medical kit. Not only can someone with allergies require them, but you'll also need them on hand in case someone gets stung or bitten by a bug. If you are unable to locate the cause of an insect bite or sting, you should clean the afflicted region and then apply antihistamines. Whatever procedure works should be used to remove any remaining stinger.

Some common bites and stings are included here, along with information on how to treat them.

- Ants. Ants have teeth. A fire ant bite causes excruciating agony and leaves tiny, transparent blisters in its wake. Cleaning an ant bite with soap and water is recommended. If necessary, provide an antihistamine. Use the bee sting kit if the individual who was bitten is allergic.

- Wasps and bees. If a wasp or bee stings you, you should use a knife or fingernail to scrape the stinger from your skin. To reduce itching, apply a cool mud mixture or a cold compress. Avoid picking at the region that stung. It is now more vulnerable to infection. If there is a bee sting kit in your medical bag, make sure you know how to use it before you leave in case someone in your group has a bee or wasp sting allergy.

- Centipedes or millipedes? Were you aware that centipedes use their front legs to inject venom? The poisons that millipedes carry on their bodies might irritate your skin. If a millipede or centipede bites you, the affected area may become red, hurt, and swollen. Use soap and water to clean the area. Apply analgesics if required.

- flies and mosquitoes. Use insect repellent to avoid getting bitten by flies and mosquitoes. To stop these insects from biting, you should also be careful to cover any exposed body parts with soil or clothes. Applying an anti-itch lotion will help lessen irritation if you are bitten.

- Scorpions. Scorpions have eight legs, a tail, and a flat, slender body with claws like lobsters. The stinger on the tail is poisonous. Although a scorpion's sting hurts, it does not kill humans. A scorpion's venom is neurotoxic. As directed in the event of a neurotoxic snake bite.

- Spiders. The majority of spiders in North America are not toxic. After cleansing the bite, you can use an antihistamine, anti-itch cream, and/or painkiller, if necessary. Having said that, you should be aware of two poisonous spiders: the brown recluse and the black widow.

- A tick. To extract a tick, grasp the base of its body and push backward until the tick comes loose. Apply an antibiotic lotion and bandage the head if it is not removed with the body. It has to be handled the same as any other open wound.

Treating Animal and Snake Bites

Most snake bites are not toxic. Poisonous ones are seldom fatal or even severely incapacitating. But you should always treat a snake bite as if it were toxic, even if you are not sure.

Two categories exist for bites from poisonous snakes. They can be either neurotoxic or hemotoxic, which means they harm blood vessels and result in bleeding. The nerve centers that regulate breathing and the heart can become paralyzed by a neurotoxic snake bite.

A bite from an animal has to be treated like any other open wound. Make sure the bite is clean. After applying antibiotic ointment, cover the wound with gauze. Apply analgesics if necessary.

Hemotoxic Snake Bites

Hemotoxic snake bites can be caused by sidewinders, sand vipers, horned vipers, puff adders, and rattlesnakes. You will experience burning at the bite location and observe one or more fang marks right away. There can be mild to severe edema after five to ten minutes. You can feel numbness and tingling in your lips, pulse, fingers, toes, and scalp within 30 to 60 minutes.

After 30 to 90 minutes, twitching of the lips, eyes, face, and neck may happen. One to two hours following the bite, symptoms including sweating, nausea, vomiting, chest tightness, elevated heart rate, fast breathing, palpitations, chills, disorientation, headache, and fainting are frequently experienced. Within two to three hours, the bite site may seem bruised and may develop huge blood blisters. After six to twelve hours, breathing difficulties, internal bleeding, and collapse are common.

Neurotoxic Snake Bites

The following snakes may bite humans: cobras, mambas, kraits, and coral snakes. After getting bitten, you could not feel anything at all. There could be some swelling and bruises, as well as some burning. However, within 90 minutes of the bite, numbness and paralysis of the affected extremities frequently develop. Increased salivation, drooling, twitching, uneasiness, tiredness, and even giddiness are typical after one to three hours. Five to ten hours following the bite, symptoms include breathing difficulties, blurry vision, and difficulty speaking and swallowing. You may not be able to survive without medical help.

Treating a Poisonous Snake Bite

As quickly as possible, anyone bitten by a deadly snake has to receive medical help. It will be necessary to take the bitten person to a medical institution. You may only provide short-term care for a dangerous snake bite victim before to and during their transfer to a hospital.

In order to cure a toxic snake bite, the person should be made to lie down and remain still. The venom will spread more widely due to activity.

Kill the snake if you can so you can take it to the doctor to be identified, but make sure you avoid becoming poisoned by chopping off and burying the head.

Employing a mechanical suction device, extract the poison from the bite. It can also be squeezed for a half hour. Avoid cutting and sucking. Furthermore, avoid using ice or spilling alcohol on the bite location.

Take off any jewelry or constricting apparel from the biting location. After cleaning the area, cover the biting location with a bandage. In case the bite is located on a limb, use a pressure dressing. To assist block the spread of poison, place it between the bite and the heart, two inches above the bite. Additionally, the extremity should be splinted and kept below the level of the heart.

Ultimate Wilderness Gear

If you find yourself in a wilderness environment, you must arm yourself to the teeth in order to tame Mother Nature. However, attempting to survive in the outdoors calls for far more than simply standard household items.

You require certain objects that will enable you to remain alive, secure, and comfortable. It might not always be feasible to fit everything into a lightweight, compact pack, though. As such, you may have to narrow down your list of essentials to a small number of goods that will enable you to survive for weeks in the wilderness.

All these goods are useful for setting up camp or establishing confidence, whether or not your outdoor expedition goes awry.

Fire Starter

Nothing is more essential than food, drink, warmth, and rest, in accordance with Maslow's Hierarchy of Needs. Perhaps the most crucial survival item you can include in your bag is a fire starter as it will enable you to boil water, make meals, remain warm, and sleep well without having to worry about predators.

There are numerous varieties of fire starters available, but the Firebiner—a carabiner with a tiny blade and a farro sparking fire starter—is our top pick due to its sheer usefulness.

Water

Water makes up more than 70% of a person's weight. Anything that throws this equilibrium off might lead to extremely dehydrated. Dehydration may be lethal in the outdoors very rapidly due to meteorological circumstances. You thus require a means of making up for any water loss from your body.

During your time in the bush, figure up a method to stay sufficiently hydrated at all times. You can only carry so much water with you at a time if you use a hydration kit or bottle. On the other hand, a water filtration system may transform tainted or dangerous water into pure, safe drinking water.

Knife

Selecting the ideal survival knife, multi-tool, or combination of the two is a very personal choice. Choose a tool that fits the size and weight requirements of your kit and has characteristics that complement your abilities and the kinds of activities you want to undertake.

While some people stick to carrying only a multitool that won't draw notice, others feel safer knowing they always have a knife with them. You may cut firewood with the use of small, machete-like blades like the Karen Hood Chopper.

Topographic Map

You should always carry a topographic map with you, even if you're heading out on a short, level day trip in a well-known area. Even in your most used stomping grounds, it's possible to become lost and turned around if you stray from the track.

Even while you won't likely get lost on a trek or backpacking trip, it's a good idea to become familiar with topographic maps. It'll be an enjoyable method to become more knowledgeable about the region and may even prove to be a life-saving survival gear.

Compass

A compass, like your topographic map, is an absolute must. If you can read an analog one correctly, it's lightweight, simple to use alone or in conjunction with other instruments. Enroll in an orienteering course offered by your nearby hiking/mountaineering association or outfitter to ensure that you are well-versed in the usage of this traditional survival technique in an emergency.

First Aid Kit

Being hurt in a survival situation is something you should be sufficiently ready for. The isolation of the environment may prevent you from reaching the emergency medical personnel in time. In this instance, a first aid kit is essential. It can assist move wounded limbs, stop minor injury bleeding from getting worse, and even dress wounds to halt infections.

You don't have to get the enormous, fully functional first aid kits. Smaller versions are useful for survivalists. Even lighter kits can be made by include only the necessities, such bandages, gauze, cotton wool, latex gloves, over-the-counter pain relievers, antibiotics, and alcohol-based cleaning wipes.

Emergency Survival Whistle

A high-quality, loud whistle is useful for many different outdoor activities. When you routinely travel into the wilderness, where there are few humans and lots of animals, it's one of the most important survival gear.

A whistle with a minimum volume of 100 dB may be heard over surrounding noise, cover large areas, and warn large animals—like bears—of your approach. If you can, keep it fastened to both your person and your bag in case you lose track of your belongings.

Rope

Everyone will elaborate on the significance of traveling with a rope, from avid hikers to the hobbits in Lord of the Rings. However, the paracord of today's survivalists—which was created in 1935 for military purposes—is the best option.

Rain jacket or raincoat

It is difficult to predict when calamity will rear its ugly head. You can't even predict the weather at that particular time. Planning for the survival journey would make things lot easier because you could rely on the department of meteorology for information.

If not, make sure you bring a raincoat or jacket so you have something to sit on when it pours. Pack a raincoat or jacket even if the meteorological department provides you with weather focus for the future, as you cannot be certain that the focus will be accurate.

Insect repellent

Our comfort level is often restricted by insects, particularly when we're out in the woods. Their propensity to transmit infections and diseases makes them even more hazardous. In order to survive in the outdoors, one of the most important items you need is bug repellant.

Eco-friendly bug repellant is what's needed. Therefore, you must locate a natural product that is effective in keeping most insects away. Fortunately, tsetse flies and other harmful insects may be repelled by natural repellents like lemon juice. To find out more, do some study.

Flashlight / Head Lamp

Having to navigate in the dark is the only thing more difficult than becoming lost, becoming stranded in a rural emergency, or attempting to survive the unexpected. It's usually a good idea to have a tiny pocket-sized flashlight or your headlamp, even if you're just planning a day trek.

Tarp

It's not always possible to construct a survival igloo, and there are instances when you should preserve your space blanket to wrap around your body rather than place it over your head. Tarps are essential survival gear because of this.

The tarp may be used in a variety of ways, such as a ground pad for your tent, a campground cover, an area to prepare meals, shielding equipment from the weather, and even gathering drinking water. Few things in any survival pack are as useful as a high-quality, durable tarp.

Signal Mirror

These indestructible reflectors, often known as rescue mirrors, are made especially for use outdoors and as survival gear. You may apply sunscreen spotlessly on any typical day by using the mirror to guide you. On the other hand, you may signal for assistance using the mirror if your voyage goes awry. Depending on the weather, the reflected gleam can travel up to seven miles before losing signal and informing aircraft of your location and state of emergency.

SPOT locator

A compass and topo map are useful tools, but having access to contemporary technologies is preferable. With a Spot, you may instantly send emergency personnel to your precise position.

No matter how far away you are, a personal locator beacon will convey your GPS coordinates and an SOS signal without the use of a mobile phone service. Some may also send messages that go beyond the standard SOS, allowing you to provide friends and family with extra information, such as the fact that you're running behind schedule but not in imminent danger.

A dry bag or Ursack

Finally, but just as importantly, a dry bag or smell-proof Ursack is an excellent method to store vital survival supplies like your phone, map, matches, firestarter, and so on. By taking the Ursack path, you may also avoid drawing unwelcome attention to your food hoard or cannabis stash as you work to return to safety.

This is a comprehensive list of everything you will need to survive in the woods. It does not imply that you won't require other things. As part of your outdoor readiness, these are the essentials you should make sure are in your pack. However, the list would be incomplete if it did not include your cerebral abilities and a trustworthy companion. In an emergency, no one can help you more than a reliable buddy. Additionally, you must be analytical, innovative, and creative in order to come up with novel strategies for surviving various crises.

Conclusion

Although becoming lost in the woods might be terrifying, you can escape a tight spot and return to safety by remaining composed and following the appropriate procedures. While you wait for help, don't forget to pause, evaluate the situation, and attempt to locate yourself. You can escape the woods and return to society with a little luck and thoughtful preparation.

As we get to the end of our outdoor adventure, keep in mind that survival is an ongoing learning process rather than just getting by. Nature has taught us that obstacles are opportunities for growth. Now that you have these newfound abilities and a stronger bond with nature, you may confidently take on the unknown. Thus, learn to live with uncertainty, have faith in your skills, and follow the spirit of the wilderness.

BONUS

Survive the Wild: 50+ Survival Tips & Bushcraft Skills

BOOK 2

SUSTAINABLE LIVING: REDUCING WASTE AND SAVING YOUR WALLET

In the hectic pace of contemporary life, where every choice seems to have an impact on the health of the planet, "Sustainable Living" appears as a beacon of hope for a peaceful coexistence. We set out on a life-changing journey in the pages that follow, one that goes beyond simple survival and into the realm of conscious living. We find a road map for a richer, more satisfying life as well as a path to a future that is more ecologically friendly as we investigate the ideas and practices of sustainable living. Come along as I untangle the strands of this story, where every chapter serves as a wake-up call for us to reconsider how we interact with the planet and how we should fulfill our responsibility as its stewards. "Sustainable Living" is more than just a book—it's a call to action and a celebration of the little decisions that have the potential to leave a lasting legacy for future generations. Together, let's set out on this journey to a world where coexisting peacefully with nature is a way of life rather than merely a choice, where every page we turn will bring us closer to that goal.

What is Sustainable Living?

Whatever name you give it, this way of life is beneficial. It's also referred to as eco-friendly living or sustainable living. A useful and environmentally friendly idea is sustainable living. It enables us to make constructive adjustments that lessen our influence on the environment.

These beneficial adjustments lessen or balance the harm humans do to the environment. Many environmental problems are caused by humans, frequently without their awareness. This claim is supported by a wealth of convincing evidence, although being hotly contested.

The Intergovernmental Panel on Climate Change estimates that about all of the warming seen since 1950 is the result of human emissions and activity. Our earth is changing due to aerosols, greenhouse gasses, and other human-caused factors.

We can shift that by leading sustainable lives. This is a style of living that considers the future of our wonderful Mother Earth. It's a method of living that will leave a more beautiful, braver, healthier, and better legacy for our kids and grandkids than the ones left by previous generations.

Another word we hear a lot is zero-waste. What distinguishes zero-waste living from sustainable living, then? Although there may be some overlap and similarities between these phrases, their meanings are distinct.

Living sustainably entails using a variety of strategies to reduce our total environmental effect. The aim of zero-waste living is slightly different. Living a zero-waste lifestyle entails minimizing the waste you generate, especially with regard to single-use plastics.

Why Should You Learn How To Be More Sustainable?

First off, why is it crucial that individuals begin incorporating some of these sustainable living suggestions into their daily lives?

Maintaining the human population requires a lot of work.

The official population of the world is 8 billion, and by 2050, the UN predicts that number will rise to 9.7 billion. If this happened, the natural resources required to support the existing way of life for humans would need the equivalent of nearly three planets.

The issue is that there aren't three Earths. We've got one.

For survival, the human population depends on both "infinite" renewable resources and limited, non-renewable resources.

We utilize quotes because, although renewable resources are theoretically limitless, they only yield a certain quantity of resources over a specific length of time. According to the Global Footprint Network, humans are depleting these resources nearly twice as quickly as the planet can replenish them.

The alarming (and ever-growing) rates at which our natural resources are being drained are unsustainable, as future generations will not be able to keep them up to date.

To exacerbate the situation, we pollute in addition to consuming.

The United States alone sent 67 million tons of pollutants into the atmosphere in 2021, according to the EPA. One of the biggest threats to human health that we currently face is air pollution. Seven million people die each year as a result of respiratory infections, heart attacks, lung cancer, and other conditions brought on by high pollution exposure.

Not only are physical but also chemical impurities clogging up streams. Two million tons of sewage and agricultural and industrial waste are dumped into our water system each day. Millions of species that depend on these priceless water sources are being wiped out in addition to the quality of our own water being diminished.

Based on estimates of 1,000–10,000 times greater extinction rates than natural ones, scientists refer to the current extinction event as the sixth mass extinction. In contrast to other extinctions, human action alone is the cause of the present one.

Climate change, water pollution, air pollution, and biodiversity loss are just a few of the serious (and perhaps permanent) environmental degradation issues brought on by our present consumption and polluting patterns.

We alone are the source of our environmental issues and the only ones who can fix them.

We can jointly halve greenhouse gas emissions by 2050 if everyone follows just a few sustainable living guidelines today. This will drive adjustments by the larger firms that account for 71% of world emissions.

Little acts add up to significant change when it comes to sustainable living, and the moment to act is now.

How to Practice Sustainable Living

The term "sustainable living" can refer to several distinct concepts. A few modifications are easy. Some are a little trickier. Here, we'll begin with the easier ones. Reading up on sustainable living is a great place to start if you're interested in doing so!

The five Rs will direct you to your next destination. To refresh your memory or if you're not aware, the five Rs stand for refuse, reduce, reuse, rot, and recycle. There's a reason some of these seem somewhat familiar to you.

Reduction, reuse, and recycling—the original three Rs—have been practiced for many years. In 1976, the Resource Conservation and Recovery Act was approved by Congress. Waste became a major issue, thus this legislation was created to encourage recycling and conservation activities.

The phrase "reduce, reuse, recycle" is said to have been coined about the same period. Since then, it has grown to incorporate two other strategies to lessen our influence on the environment: decay and trash.

Reuse, recycle, and reduce are very self-explanatory concepts. Since the majority of us heard them frequently as children, they have been engrained in our daily lives. Fortunately, they are also some of the simplest approaches to lead a more sustainable life.

Reduce

Reducing is merely surviving with less. And for the most part, that is true. Do you let the water flow as you brush your teeth? Even when there is a plenty of natural light coming in via your windows, are the lights on?

Reusing extends the life of commonplace items. This might be anything from purchasing items at secondhand stores to give it a second chance at life to using reusable tote bags for your supermarket shopping.

Every year, two billion single-use plastic razors are discarded. On the other hand, reusable ones provide a closer shave, don't wind up in a landfill, and eventually save you money.

We may repurpose objects such as discarded coffee cans, pasta jars, candles, and baby food containers in addition to commonplace cosmetic products like these. To make these things look more ornamental, you can wrap extra wrapping or scrapbook paper over them.

After that, you may utilize them to store anything in your kitchen, bathroom, or home office. Use them in any creative method that comes to mind. For example, put all of your pencils in an empty container, paper clips in a baby food jar, hair ties in an empty candle jar. Be imaginative!

Finally, we all understand what recycling entails. Sadly, recycling is no longer sufficient because only 9% of materials designated for recycling are actually recycled. Even with their labels, many plastic objects are just too hard to recycle.

We thus also require the additional Rs. What to do with all the plastics in our life is not the important thing. The secret is to not use plastic in order to stop more of it from being made.

Refuse

Refusing is comparable to cutting. Refusing, however, entails preventing the addition of new objects rather than cutting back on things like water and utility consumption. Instead of putting stuff in your virtual shopping basket, take a time to inventory what's in your closet.

Do you need any more sweaters and yoga pants? Or did you just purchase on a whim and already had everything you needed? It's crucial to understand that we are not attempting to intimidate anyone into not purchasing necessities.

All we're doing is educating our aware clientele that necessity and want are not the same thing. When purchasing purchases, make an effort to be more thoughtful and consider whether they are really essential. This saves money, lessens the amount of stuff we don't need, and benefits the environment.

It's crucial to enforce the reject rule for items like water bottles and single-use plastic bags. The degradation of these objects may take up to 500 years. Furthermore, they will soon become outdated given the abundance of modern alternatives.

Rot

The term "rot" carries a bad connotation. However, rot is really a wonderful thing in an eco-friendly environment since it indicates that something is properly returning to the soil. Composting is the method used to accomplish this decomposing.

What is compostable?

The process of composting involves adding organic components to the soil. It promotes the growth of beneficial bacteria and fungus, lessens the need for fertilizers, enriches the soil it penetrates, aids in the suppression of pests and plant diseases, and helps plants flourish.

To your surprise, we can also compost yard trash and food waste. We distribute a lot of our items in packaging or wrapping that is biodegradable and produces no trash.

Eco-friendly living practices

Refusing, reducing, reusing, decomposing, and recycling are some of the most effective strategies to get started on your environmentally responsible path. Although we provided a few examples in the preceding sections, there are many more approaches to sustainability.

Here are a few more instances of living sustainably:

Make the switch to LED light bulbs, which have a lifespan of around 50 times longer than typical incandescent ones.

Reduce your meat consumption, particularly red meat, to help reduce greenhouse gas emissions.

Whenever possible, go for reusable products.

Make the switch to electronic documents and dishcloths from printed pages and paper towels to become paperless in your home office and kitchen.

Grown your own food in the garden (or, if you don't have much room, in hanging flower boxes).

Instead of tossing away unwanted kitchenware, clothes, and other stuff, donate or give them away.

When possible, use a clothesline to dry your clothing.

Instead of standard product containers packed with chemicals and hefty plastic packaging, choose eco-friendly soaps and laundry detergents.

Apply shampoo and conditioner bars in the same way.

While you're doing the dishes, switch off the faucet and take shorter showers.

To help cut down on greenhouse gas emissions and, in the case of some of these strategies, improve your health, walk, cycle, take the bus, or carpool.

To prevent food waste, freeze or consume leftovers.

Try purchasing secondhand or trading with friends to make your closet more sustainable.

To reduce chemical pollution, purchase or produce environmentally friendly cleaning products.

Making Greater Progress Towards An Eco-Friendly Lifestyle

Every item on our previously mentioned list will assist you in leading a more sustainable lifestyle. The majority of them were simple, little actions that may have a significant impact. However, when we're prepared, we may also take on more ambitious tasks.

The following are a few of the largest and most important ways we may transition to an environmentally friendly style of living:

Installing solar panels.

replacing drafty, outdated windows with more energy-efficient models.

Energy-efficient equipment, such as dishwashers, washers and dryers, and refrigerators, should be substituted with damaged ones.

Install faucets and showerheads that conserve water.

However, don't allow these larger objects deter you. You may also take a ton of free, smaller, more cheap steps, as well as many in between.

Does it Cost a Lot to Live More Sustainably?

It's not necessary to spend a fortune on sustainable living. Even though they may initially cost a little more, reusable items nearly always result in long-term cost savings. Furthermore, even the larger actions that we previously outlined may be divided into smaller, more doable ones.

Start small with solar lights or lamps if you're not ready to install solar panels on your roof, for instance. They save electricity and money by charging in the sun during the day and then emitting enough light to keep the porch lights off.

Utilize them to adorn your backyard, driveway, and front path. Play around with the many shapes and sizes of solar lights! And living sustainably on a tight budget is just getting started. There are a ton of other affordable and impactful ways to make a difference.

Budget-friendly sustainability tips

Many people have the idea that living sustainably costs more money. This need not be the case. Numerous green living tips are inexpensive, sometimes even free when compared to their conventional counterparts.

Here are a few of our preferred low-cost eco-friendly living strategies:

Rather than adjusting the thermostat once again, grab a blanket. (When you are at home and awake, the Department of Energy advises keeping the thermostat at 68 degrees.)

To begin a compost pile, fill a container with materials such as coffee grounds, leaves, shredded paper, wilting vegetables, and grass clippings.

Use cold water to wash your garments rather than heated or hot.

Take advantage of LED light bulbs available at the $1 shop—yes, the dollar store!

Take reusable shopping bags with you whenever you go shopping.

A great deal of what we've included in this part and the sections above is really affordable. Some even help you save money, such as buying used clothing, consuming less water, walking or bicycling to work, and limiting your meat intake.

Waste reduction strategies

There are numerous, little methods to cut waste that match your lifestyle, from making eco-friendly gift choices to adopting sustainable behaviors. We're giving a list of 21 simple zero waste recommendations to get you started. There's no judgment or pressure; simply try a couple and discover which one suits you best!

And never forget: excellence should not be the adversary of goodness. Millions of ecologically conscientious individuals performing flawed zero-waste practices are what the planet needs, not a select few doing it flawlessly.

The best environmentally friendly items are bulk commodities.

Think about purchasing necessities in large quantities and keeping them in reusable containers like canvas bags or mason jars. You'll be able to obtain just what you want because you have control over the quantity. Pre-packaged items typically have a high price tag, so as an extra bonus, you should see a change in your bank account. Consider joining or starting a grocery buying club to save even more money!

Skip the plastic bottles.

Tap water makes up to 40% of the bottled water that is sold. Invest in a high-quality water filter and a cute reusable water bottle rather than falling for deceptive marketing. There are many more sustainable solutions besides glass, which we adore. Get a thermos for your tea or coffee to go while you're at it. You'll be able to customize it to your exact specifications and save money in the process.

Purchase a set of fabric produce bags.

Do you recall those thin plastic bags they sell in the vegetable section? excessive waste. Purchasing a set of cloth produce bags and being sure to pack them can safeguard your purchases and stop further plastic from finding its way into the ocean and the waterways that support aquatic life. Additionally, these bags work well for carrying big items. You'll never find produce cuter!

Keep fruit free of plastic wrappers.

This is a difficult one, as many grocery stores shrink-wrap every loose produce item they can find, including some with naturally occurring biodegradable packaging. Bananas covered in shrink wrap, anyone? Which takes us to our next piece of advice!

Encourage your nearby farmer.

The food will be far more nutrient-dense, fresher, and tasty as well as more sustainable. Additionally, by buying locally grown food, you'll be reinvesting your money in your community's economy. As an alternative, you might register with a nearby meal delivery business or a csa.

Don't use single-use plastics.

Straws, plates, glasses, cutlery, and other plastic items clutter landfills and wind up in sea turtles' tummies. Thus, when you're out and about, stock up on reusable containers, stainless steel straws, and silverware to carry in your luggage or vehicle.

Vertebrate your closet.

Shop only at sustainable clothing manufacturers or visit your neighborhood consignment or secondhand store. Unique, gently used, or even brand-new items are frequently available at a small portion of their original cost. acquire used to the question "where did you acquire that?" from folks. Donate the pair of pants you haven't been able to fit into for ten years while you're there. According to Marie Kondo, "you have to get rid of items that are no longer useful in order to genuinely treasure the things that are significant to you.

Arrange your food.

Because meal planning guarantees that you'll consume everything you buy, it may actually help you cut expenditures on food, simplify your grocery visits, and minimize waste. It will also spare you from having to answer the question "what's for supper tonight?" which almost everyone dreads asking after a demanding workday. Who knows, too? A little extra "you" time on hectic weeknights may be yours.

Allow vegetables to shine.

In an ideal zero-waste society, industrial farming's detrimental effects on the environment would be eradicated and everyone would adopt a vegan diet. However, if you can't or won't give up meat, go for locally farmed meat. And think about starting a new custom at your home: meatless Monday. You may find that plant-based meals may be surprisingly tasty and fulfilling.

Recycle your leftover food.

Yard debris and food scraps, which may be composted instead, make about 30% of the garbage that Americans create, according to the Environmental Protection Agency. We may lower the amount of landfills and the emission of methane, a greenhouse gas that is a key contributor to global warming, by redirecting that garbage away from the dump. Commencing is a simple task. You have two options: start your own compost pile or use the city's compost pickup service.

Use bees wrap instead of plastic wrap.

They are the playful, environmentally friendly relative of plastic wrap, made of beeswax, fabric, oil, and pine resin. There are many different sizes, shapes, and designs available, so you're sure to find one you enjoy. They save an incredible amount of plastic trash, are reusable, and function remarkably well. They may also make wonderful gifts and are rather simple to create!

Utilize vintage clothing

For serious cleaning tasks around the house, make fabric rags from of old, undonateable garments and use them in place of paper towels. whether you frequently go through a lot, see whether your neighborhood thrift store sells large quantities of rags made from unsaleable contributions. Alternatively, if a t-shirt just no longer fits, make a reusable bag out of it!

Produce your own cleaning supplies.

It may surprise you to learn that many traditional cleaning solutions include hazardous chemicals and endocrine disruptors that seep into our bodies, rivers, and sewage systems. Thankfully, creating environmentally friendly cleaning solutions at home is not too difficult. For example, you can clean your floors, countertops, showers, and toilets by mixing vinegar, lemon juice, and baking soda.

Sustainability Tips For Families

Make the switch to eco-friendly diapers or reusable cloths. Disposable diapers make up around 4% of solid waste in landfills, making them the third most common single consumer item.

If you are able, decide to breastfeed. Four months of exclusive formula feeding has a 35% to 72% greater carbon footprint and environmental effect than four months of exclusive breastfeeding. Breastfeeding is the most sustainable method of child rearing since it requires no packing, heating, or transportation.

Give up using plastic bottles. Glass bottles are superior than plastic bottles when it comes to infant bottles since they are easy to reuse and do not leak pollutants.

Prepare baby meals at home. You may be in charge of the ingredients and quantity of plastic you use when you supplement your baby's diet with homemade food, but make sure you speak with their physician to make sure you understand their needs.

Make an eco-friendly toy purchase. According to research, dangerous substances that are poisonous to kids and the environment are present in 25% of children's toys. For everyone concerned, those constructed of natural materials like bamboo or wood are far superior.

Eliminate baby wipes. Wet wipes are frequently consumed by marine life searching for their next meal since they do not break down before entering the sewer. Eco-friendly baby wipes made from biomaterials have 38% less of an environmental effect than those made from petroleum.

Introduce children to sustainability at a young age. Teaching our children to take responsibility for the amazing planet we live in is one of the most sustainable things you can do at home. For children, promoting sustainability may be as simple as reading The Lorax before bed or as involved as enlisting their assistance in the garden.

Sustainability Tips For Pet-Owners

Make use of green kitty litter. Cat litter made of wheat, walnut, or wood is a fantastic substitute for clay or silica-based litter, which is not biodegradable.

Select pet toys that are sustainable. Safer alternatives to plastic dog toys are eco-friendly ones made of natural materials like rice husk, rubber, and hemp, or recycled materials like used t-shirts.

Gather rubbish and place it in biodegradable bags. Unlike their plastic equivalents, they may disintegrate in water in as little as three months, so they won't accumulate in landfills.

"Swap Don't Shop." 920,000 of the 6.3 million companion animals that enter American animal shelters each year are put to death, according to the ASPCA. Adopting rather than reproducing reduces the amount of resources required to provide food for another animal.

Bring pets inside. Because they are opportunistic predators, outdoor cats pose a serious risk to the biodiversity of the local fauna. Every year, household cats that roam freely kill up to 4 billion birds and 22.3 billion animals.

Neuter or spay your pets. Important population control techniques include spaying and neutering, which lower the amount of resources required to maintain the population.

Look for environmentally friendly cat food and sustainable dog food. This can lessen the 106 million tons of CO_2 that are released into the atmosphere during the manufacture of pet food.

Conclusion

At first, sustainable living may seem a little overwhelming. However, you quickly discover how simple it is to put these tiny adjustments into practice—which, when done one small step at a time, actually have a big influence on your surroundings. All we need to do is practice more aware living and abstain from actions that we know may be detrimental to the environment. Living a more sustainable lifestyle increases your awareness of your surroundings, the things you buy, eat, and the potential trash you produce. This increases your empathy for the environment and your knowledge of how you can protect it.

BONUS

SUSTAINABLE LIVING ON A BUDGET: 10 eco friendly tips to save money while saving the planet

Book 3

Raising Livestock:
A Comprehensive Guide to
Animal and Poultry Breeding

Although farming with animals is common and may appear to be the best option for novice farmers, you will fail if you don't do your studies.

One of the main agricultural sectors in the nation is livestock farming, with the most common types being beef, chicken, pig, and sheep. Although the majority of these species are farmed nationwide, not all habitats and locations are suitable for all kinds of animals. An animal's ability to survive in its surroundings has a direct impact on its production and overall health.

Healthy animals are a farmer's best source of income. So keep in mind that an unwell animal won't truly be producing anything since it will use all of its energy to become healthy again. It will thus not produce a calf for you, develop in a feedlot, and not lay an egg if it is a chicken. In order for an animal to be used for production, we must ensure that its physiological systems are maintained and that any extra energy is directed toward this purpose.

Animals grown or kept primarily for their meat, milk, and meat products are referred to as farm animals. in addition to earning a living and supporting agricultural operations.

Animals raised on farms for meat, dairy products, or to help farmers are known as farm animals. "Livestock" is one of the more often used names to describe them. Farm animals are not the same as wild animals. They are able to coexist. But in order to survive, wild animals must be kept far from people.

Any animal reared or bred on a farm is referred to as a farm animal. The goat, sheep, cow, camel, buffalo, and several types of horses and donkeys are a some of the most popular ones.

Pigs and an assortment of fowl, such as chicken, turkey, duck, and geese, are also present. Among other species, the category of micro-livestock includes rabbits, guinea pigs, and cane rats. There are too many creatures to name that may reside on a farm. Here are a few of the more common ones, along with some details on their uses and how they help people.

Moving out of the city is just a pipe dream for a lot of individuals. Others might consider their tiny farm to be more feasible. Whichever kind of prospective farmer you are, choosing to include livestock animals in your plans is a significant choice.

Having animals is frequently the epitome of the back-to-the-soil ideal. It might be a fruitful endeavor. The delight of stocking your freezer and refrigerator with meat and eggs from your own livestock is hard to dispute.

Naturally, having cattle involves much more than just giving them food and water and maybe making an occasional trip to the veterinarian. Taking care of cattle is a full-time, rewarding, and risky profession. Set objectives for this next phase of your trip before making an animal investment. To determine whether the animals you're interested in are a suitable fit for your farm, find out more about them.

Things to Consider Before Raising Livestock

What Size of Land Do I Need?

Even with just one animal, growing cattle takes up more area than rearing hens or rabbits. Consider the following: how much land do I now own and how much will I require to raise various animals?

There's a lot to think about. Will the animals be given exclusively grass or mixed grain and grass? How many cattle could you keep on, say, an acre of grass if they were only fed grass? The number of sheep?

More cattle may be kept on one acre for a shorter amount of time by rotating them. The amount of grass-fed cattle you can sustain will be constrained if you are unable to rotate them. It's possible that your property cannot sustain more than a small number of beef cattle. On the other hand, you could have enough space to rear fifty or more hens.

What Local Laws Apply to Me in My Area?

Speaking of animals, you should familiarize yourself with any local regulations on livestock ownership. It could be necessary for you to have particular licenses or permissions if you intend to breed your animals or sell the items they produce.

In a similar vein, you must ascertain if your property is allocated for farming or animals. To get the answer, just pay a brief visit to your town office and request a local zoning map.

Is It Within My Budget to Keep a Herd?

Yes, there will be early start-up costs. These may include purchasing building supplies, livestock and feed, as well as paying for permits and licenses. However, it doesn't stop there. Veterinary care, fresh water, feed, and other expenses may mount up rapidly. Equipment breaks down and animals become ill. Should an illness wipe out your herd, would you be able to purchase replacement animals?

Do not forget that markets change if you want to sell your dividend. Do you possess the necessary funds to maintain your business in the event that the price of milk or meat drops? It's critical to have a realistic outlook on your money. Although it may be possible to own animals, there are realities that come with that goal that you must take into consideration.

Am I Prepared to Manage Predators and Pests?

Undoubtedly, your pals may drop over to check how you're getting along, and your in-laws may wonder why you're together. However, so will other, less invited guests. Do you still recall the fox you spotted across the street in the field? Or the hawk perched in the tree, its claws gripping a squirrel?

Other animals are drawn to animals. Predators are going to ultimately become interested in your new flock of chickens. It will be your responsibility to keep your animals safe, and it will take ongoing attention.

You may draw skunks, possums, and other scavengers in addition to predators. Your feed shops and trash will draw these creatures.

Will My Free Time Be Given Up?

Are you a happy worker? Is it okay if I get my hands dirty? Do you value animals so highly that you would give up the majority of your leisure time to them?

If needed, you might be able to hire a livestock sitter to help with feeding, watering, and even milking your animals. What if you are unable to?

Cows are not day-trippers. The horse will not wait to foal till the conclusion of the NFL season. Even if it's 95° outdoors, are you prepared to fix the electric fence in the middle of July?

Having animals is a full-time commitment that lasts for 365 days a year. It's challenging to travel with ivestock. That three-day music event you've been dreaming about since last year might not be possible for you to attend. The work will determine your fate.

Naturally, not every animal needs the same level of care. But generally speaking, living with cattle means not taking spring break, summer vacation, or business holidays.

Types of Livestock To Consider

Cattle

Cattle farming for beef production is a great choice for animals. They are reasonably easy to keep and can provide both meat and milk. Despite doubts about the health advantages of beef, there is still a strong market for organic, grass-fed beef.

Land needed for cattle might be substantial, dependent on the size of your herd. Approximately two heads of cattle may be fed 1 acre of grass each year. It's best to rotate your animals if you have too many of them. This will enable the pastures to recuperate, which would need purchasing more acreage.

Intelligent structures are not necessary to maintain happy beef cattle, yet bigger herds may find it difficult to stay organized. Take into consideration ear tags as a simple way to protect children from anonymity.

Another, but more complicated, alternative is milk cows. Small herds may find it challenging to make a profit due to the volatility of milk prices. The situation gets much more difficult when you take into account the requirement for winter lodging and food.

Having a Jersey or two may be a great way to supply the house and friends with milk. To properly milk if you don't want to do it by hand, you'll need a costly infrastructure. Another potential supply of meat is milk cows. Milk cows could be an excellent option if you just want to own a small number of animals.

Chickens

A common sight on many homesteads and hobby farms is the omnipresent chicken. They offer wholesome eggs for the refrigerator and meat for the freezer, so this is understandable.

Chickens require minimal time or space. All it takes is a few hours a week to clean, water, feed, and gather eggs. They are also mobile and will contribute to reducing tick populations, which are an increasing problem in many parts of the nation.

There are many different breeds of chickens; some are better for laying, while others are better for meat. Simplifying the procedure for your birds may be achieved by having a well-thought-out aim.

Goats

Goats are a reliable, somewhat low-maintenance livestock choice with a variety of advantages. Approximately 70% of those who regularly eat meat choose to consume goat as their meat of choice worldwide.

Over the past ten years, goat farming has become much more popular as a means of producing meat, milk, or cheese. Goat flesh is thin. Compared to beef and chicken, it is lower in calories and lower in fat than pig and chicken.

Furthermore, both domestically and internationally, goat milk and goat cheese are becoming more and more popular specialty market items. In the US market, goat meat is less common than beef, hog, and chicken. Nonetheless, there are still ways to make money in specialized sectors. Meat from goats is acceptable for both Kosher and Halal diets. You could have a steady consumer for your meat if you can establish a relationship with the nearby mosque or temple.

Goats can be grown as breeding stock to produce more fleece, or they can be raised for the luxurious Angora and cashmere fleece. Since goats are browsers rather than grazers, sustainability is another alluring advantage of owning one. An acre of grass can support ten goats, or even more, if it can support a few beef cattle.

Sheep

Sheep have long been a valued and well-liked livestock choice, especially in the wool sector. Fleece's value has decreased over the last few years; estimates range from $0.87 to $5 per pound.

Now, feeding a sheep for a year can possibly cost more than the fleece yields. This has caused a lot of farmers to either sell their wool in the hopes of getting higher prices the next year, or to simply destroy it. However, markets change. You have to be inventive and have keen eyes if sheep are your preferred animal.

Even in specialized and religious sectors, lamb remains a popular meat. In order to produce marketable lambs in time for Christian, Jewish, and Muslim holy feasts, hair (meat) sheep can breed out of season.

In the specialized market, fleece is still valuable. Wool may cost a lot of money to those who hand-spin yarn and produce felt. But often, they are limited to wool from certain breeds or from extremely rare sheep.

Pigs

Yes, on occasion. Pork that is reared humanely and organically is still in demand, even if many commercial piggeries have closed. Pigs require little time or effort to care for, eat almost everything, and are easy to maintain.

Pork is a handy and amiable animal. They can supply enough pork for the entire year and be a wise decision for your farm.

Rabbits

Most rabbits are kept for their fur or as show animals, pets, or both. It could make sense for you to raise rabbits. A benefit of rabbits is their sustainability. growing rabbits has less of an impact on the environment than growing other animals, even if they only eat a small portion of the feed. Furthermore, because they are little, rabbits don't need a lot of area.

Rabbit meat has more protein content than chicken or beef at the same serving size, despite not being a major meat market animal. It supplies iron as well. Find out whether there's a local market for rabbit meat if you wish to grow them. If so, you might have to invest the time in building connections with nearby farmers markets or food stores.

Animal husbandry techniques

A field of agriculture dedicated to feeding, breeding, and raising animals in order to produce valuable goods. In essence, animals are raised for a variety of goods, such as:

- Meat: Cattle, sheep, goats, and other animals

- Milk: cows, buffaloes, camels, goats, and other animals
- As physical work on farms; these animals include yaks, bulls, and horses.
- For eggs: Poultry birds, such as ducks, geese, and hens

The following are some methods of animal husbandry used in livestock farming:

Choosing the Correct Breeds: Pick livestock breeds that are compatible with both the local environment and your particular farming objectives.

Take into account elements like disease resistance, climatic resilience, and the livestock's intended use (meat, milk, wool, etc.).

Appropriate Nutrition: Considering your livestock's age, breed, and intended use, create a diet that is both balanced and nutrient-rich.

Make sure there is always availability to fresh, clean water.

When necessary, add vitamins and minerals to their diet.

Comfortable Shelter: Give animals access to clean, well-ventilated shelters to shield them from inclement weather, including intense heat, low temperatures, and precipitation.

The animals' comfort and well-being are enhanced by using appropriate bedding and flooring materials.

Healthcare Procedures: Establish a routine for immunizations, parasite control, and illness prevention in your pet's veterinarian treatment.

Keep a watchful eye on the animals' health, and take quick action to handle any indications of disease or suffering.

Breeding Management: To preserve your livestock's health and genetic variety, engage in ethical breeding.

Take into account elements like the ideal breeding age, genetic characteristics, and the animals' general wellbeing.

Fencing and Containment: To safeguard pastures and keep cattle from roaming or falling victim to predators, use the proper fencing.

Systems of rotational grazing can minimize overgrazing and maximize pasture usage.

Waste Management: To deal with manure and other byproducts, put in place efficient waste management systems.

In order to promote sustainability, think about composting or recycling garbage for use as fertilizer.

Record-Keeping: Keep thorough records of every animal, including information on their breeding history, birthdates, and medical histories.

For the purpose of monitoring the cattle's general health and production, this information is invaluable.

Training and Handling: Teach cattle to obey simple orders to ease everyday activities and lessen animal discomfort.

In order to foster trust and reduce tension during interactions, treat animals with kindness and calm.

Environmental Enrichment: To keep people from being bored and to promote natural behaviors, provide them with environmental enrichment.

Playthings like scratching posts, for example, might help animals feel better mentally.

You may improve the production, health, and general well-being of your cattle while also enhancing the sustainability of your farm by implementing these strategies into your farming operations.

Breeding and care of livestock

Animal breeding is a crucial aspect of animal husbandry. Breeding is the process of selecting two animals with superior traits and then crossing them to develop the desired features. a collection of creatures that are connected by ancestry and share the majority of traits, including size, characteristics, and overall appearance.

The two main objectives of breeding are to:

- Boost animal productivity
- Enhance desirable traits in products.

Types of Breeding

Inbreeding

Inbreeding is the term used in animal husbandry to describe the mating of more closely related individuals of the same breed over four generations. Identified and mated are superior males and females of closely related individuals. Here are the results of inbreeding:

- produces homozygous pure animal lines.
- reveals recessive genes linked to undesired traits that may be changed or removed.
- results in the accumulation of advantageous genes and the eradication of harmful genes.

The animal's fertility and production decline with ongoing inbreeding. We refer to this as inbreeding depression. By mating the animal with a superior unrelated animal of the same breed, it can be defeated.

Outbreeding

The crossing of animals from various breeds is known as outbreeding. It encompasses interspecific hybridization, cross-breeding, and out-crossing.

The mating of animals of the same breed that have not shared ancestors for four to five generations is known as out-crossing. It is typically used to animals whose development and production rates are below normal. Overcoming inbreeding depression is aided by it.

The mating of a superior male from one breed with a superior female from another is known as cross-breeding. Hybrid vigor is the result of combining the best traits of both breeds. This kind of offspring is referred to as a hybrid. A hybrid can be kept in its current form or further exposed to inbreeding. For instance, Hisardale sheep are a cross between Marino rams and Bikaneri ewes.

Males and females of distinct but related species mate in a process known as interspecific hybridization. Progeny has traits that are valuable to both species. An interspecific cross of a donkey and a horse is a mule. It is more resilient to illness than a horse and faster and stronger than donkeys.

Controlled Breeding Techniques

1. Artificial Insemination

The male produces sperm, which is then injected into the female's reproductive system. Semen can be utilized right away or preserved for later use.

2. Multiple Ovulation Embryo Transfer (MOET) Technology

An FSH-like hormone is given to the cow, causing superovulation and follicular maturity. During superovulation, six to eight eggs are generated every cycle as opposed to just one. The cow might be artificially inseminated or paired naturally with a better bull. A fertilized egg is non-surgically retrieved at 8–32 cell stages and given to a surrogate mother.

Once more, the genetic mother is open to undergoing another superovulation cycle. This method has been successfully used to breed lean meat-producing bulls and high milk-yielding breeds of females.

Disease prevention and treatment

Keeping your animals healthy and happy is essential to success in the complex dance of livestock farming. This guide offers insights and solutions to promote a robust and vibrant cattle community, acting as a compass through the challenges of disease prevention and treatment.

1. Vigilant Observation: Develop your keen observation skills to start your trip toward the health of your cattle. Keep an eye out for any small changes in your animals' appearance, behavior, or hunger. Effective illness prevention frequently depends on early identification.
2. Strategic vaccination Protocols: Create a thorough immunization schedule under the direction of a veterinarian. Immune system strength is increased by vaccinations, resulting in a strong barrier against common illnesses. Maintain thorough documentation to guarantee on-time booster injections and optimal protection.
3. Biosecurity Measures: To reduce the possibility of the entry of illness, put strong biosecurity measures into place. To protect your cattle community, quarantine recent arrivals, manage foot and vehicle traffic, and keep a secure boundary.
4. Nutritional Fortification: The cornerstone of strong health is a diet that is well-balanced. Make sure the nutrients your animals need to support their immune systems are given to

them. Adjust feeding schedules to meet their individual requirements, taking into account things like age, reproductive health, and seasonal fluctuations.

5. Hygiene and Cleanliness: Keep your livestock's living quarters immaculate. Keep living areas, feeding places, and water supplies clean and sterilised on a regular basis. One of the most effective tools for stopping the transmission of infectious organisms is a clean environment.

6. Prompt and Accurate Diagnosis: In the event of disease, prompt and precise diagnosis is critical. Develop a relationship with an experienced veterinarian who can quickly detect and treat new problems in addition to doing routine health checks.

7. Targeted Treatment Plans: The animal receiving treatment has to have it precisely catered to in order to meet its demands. Create treatment programs with a veterinarian's advice, making sure the right drugs are used at the right amounts. Observe the suggested time frames for medication discontinuation.

8. Herd Health Management: Take a comprehensive approach to livestock health by taking into account the welfare of the entire herd. To stop the spread of illness, isolate sick animals as soon as possible. You should also create strategies for the diet, immunization, and routine monitoring of your herd.

9. Never Stop Learning: Remain up to date on new medical discoveries, veterinary medical innovations, and changing best practices. Participate in workshops, interact with veterinary specialists, and establish connections with other agriculturalists to expand your understanding and modify your methods appropriately.

10. Compassionate Care: Above all, take a compassionate attitude to illness prevention and treatment. Developing a trusting relationship with your animals promotes a peaceful and robust farming community in addition to helping them heal.

You may start your road toward proactive illness prevention and compassionate treatment by adopting these techniques. In addition to demonstrating your prowess as a farmer, your dedication to the health and welfare of your cattle is a pledge to build a robust and prosperous agricultural heritage.

Conclusion

Owners of commercial cattle are aware that market conditions affect their capacity to make a profit. Markets experience ups and downs. It's a safe wager that farmers raising traditional livestock will need to adapt to shifting markets as cultural norms and tastes change.

The majority of those seeking for practical cattle ownership don't have ambitious business goals. However, if you know how to market your cattle, you may still turn a profit.

You can make money with fleece, milk, meat, cheese, and eggs by using niche and value-added marketing. Offer the flesh of your lamb or goat to the ethnic communities in the area or on holy days. You can farm animals for their byproducts rather than for their meat. These include fleece and fiber for specialist mills, and milk for cheese and yogurt.

All of your planning will pay off when the time comes for you to begin producing animals. Planning, practical thinking, and research will all contribute significantly to your final success.

BONUS

Top 3 EASY MEAT ANIMALS for BEGINNERS to RAISE AND PROCESS

Book 4

Emergency Care Essentials: Medical and First Aid for Survival

The opportunity to unplug and get some alone time away from the city is one of the attractions of outdoor adventures. It's crucial to be knowledgeable about wilderness first aid because that also implies you're further away from ambulances and urgent care centers. Consider it a chance for you to grow stronger and more independent.

Most problems that occur outside are rather small and may be handled with ease. Most of the time, your objective while administering assistance in the outdoors is to prevent a condition from getting worse so you may carry on with your expedition. Having stated that, it's critical to be ready for everything.

This is an all-inclusive reference to emergency medical procedures, assembling a first aid bag, and fundamental first aid skills. We will cover key methods and resources in this book to assist you in handling medical situations, whether you are in the backcountry or dealing with a catastrophe's aftermath.

Basic first aid skills

Assessing the Situation

Your first line of defense in the unpredictable terrain of an emergency is your ability to swiftly scan the surrounding surroundings. This ability not only keeps you safe but also enables you to make wise judgments that benefit other people.

Recognize possible hazards in the environment, such as fire, shaky structures, or dangerous items.

Pay attention to your surroundings and any potential dangers.

Put the safety of injured people, onlookers, and yourself first.

Rapid assessment techniques: Look for any threats and, if required, take precautions to safeguard the environment.

Determine the total number of participants and evaluate their health.

Determine the extent of injuries and rank the need for care in order of urgency.

Recall that the foundation for an efficient first aid response is a prompt and precise evaluation.

ABCs of First Aid

Your first approach to someone in distress should be guided by the fundamental principles of First Aid, which are Airway, Breathing, and Circulation. Comprehending these fundamentals is essential for delivering prompt aid and maybe preventing fatalities.

Airway: Make sure there are no obstacles in the path of the airway.

To clear the airway, tilt your head back a little.

Take care not to strain the neck excessively, particularly after trauma.

Breathing: Check for breathing by feeling, hearing, and seeing.

Start rescue breathing if the victim is not breathing.

As you continue to monitor breathing, provide support.

Circulation: Look for indicators of blood flow, including a pulse.

As part of CPR, start chest compressions if there is no pulse.

Wound Care

An essential skill for reducing the risk of infection and promoting the best possible recovery is proper wound care. Knowing the fundamentals of wound care is crucial for quickly and efficiently helping in emergency circumstances.

Cleaning Wounds: Before contacting the wound, properly wash your hands.

Use a sterile solution or mild soap and water to clean the wound.

Gently remove any foreign things.

Bandaging and Dressing: Select the right dressings according to the size and kind of the wound.

Tighten bandages just enough to encourage blood flow, but not too much.

While the wound heals, keep an eye out for any indications of infection.

CPR (Cardiopulmonary Resuscitation)

CPR is a life-saving procedure used to keep essential organs oxygenated and blood flowing even after the heart has stopped beating. Being able to do CPR when necessary and on time can be extremely helpful in emergency situations.

To identify cardiac arrest, make sure the patient is breathing and responding.

Start CPR right away if the person is simply gasping or not breathing at all.

Place your hands on the bottom part of the breastbone to do a chest compression.

Apply compressions at a pace of 100–120 per minute using your body weight.

Two rescue breaths should be given following 30 compressions.

Make sure your chest goes up and down with each breath.

Handling Fractures and Sprains

Fractures and sprains are typical injuries in emergencies that need to be treated carefully and quickly. Understanding how to treat sprains and stable fractures helps to both relieve pain and stop more damage from happening to the wounded.

Stabilizing Fractures: Use splints or other readily accessible materials to immobilize the damaged limb.

Padding should be used to support the damaged region to stop mobility.

For a more thorough evaluation, seek out expert medical assistance.

Taking Care of Sprains: Give the injured region R.I.C.E. (rest, ice, compression, and elevation).

Apply a compression bandage to minimize edema.

Urge the wounded individual to refrain from bearing any weight on the afflicted limb.

Gaining proficiency in these fundamental first aid techniques can enable you to react appropriately in emergency scenarios, creating a safer atmosphere for you and individuals in your vicinity. Always keep in mind that practice makes perfect, so spend some time becoming used to these abilities through frequent training and role-playing.

Emergency medical procedures

This chapter, which delves into particular medical treatments intended to address severe conditions, is the core of emergency care. Welcome. Knowing these emergency procedures can guarantee that you are prepared to handle potentially fatal situations with assurance and efficiency, ranging from burns to choking.

Choking

The Heimlich Maneuver:

- Stand behind the choking person and place your arms around their waist.
- Make a fist and position it just above the navel, thumb side in.
- Grasp the fist with your other hand and give quick, upward thrusts.

Alternative Techniques:

- For pregnant individuals or those unable to stand, perform chest thrusts.
- Encourage coughing if the person is able to produce sound.

Shock

Recognizing Shock:

- Look for signs such as paleness, rapid breathing, and weak pulse.
- Help the person lie down and elevate their legs slightly.

- Keep them warm with a blanket, and reassure them until help arrives.

Caution:

- Avoid giving the person anything to eat or drink.
- Do not raise the legs if it causes pain or discomfort.

Heat-Related Illnesses

Identifying Heatstroke:

- Look for hot, dry skin, confusion, and a rapid, strong pulse.
- Move the person to a cooler place and cool them down using water and fanning.
- Call for emergency help if symptoms persist.

Managing Heat Exhaustion:

- Move the person to a cooler area and have them rest.
- Provide cool water to drink and use cold compresses.

Hypothermia and Frostbite

Recognizing Hypothermia:

- Look for shivering, slurred speech, and weakness.
- Gradually warm the person by moving them to a warmer place and using blankets.
- Seek medical attention if needed.

Treating Frostbite:

- Gently warm the affected area using body heat or warm water.
- Do not rub the frostbitten area, and avoid direct heat.

Dealing with Burns

First Aid for Burns:

- Cool the burn under running water for at least 10 minutes.
- Cover the burn with a sterile dressing or cling film.
- Seek medical attention for severe burns or burns on sensitive areas.

Caution:

Applying cold, creams, or sticky bandages to burns is not advised.

Leave blisters alone; do not break them.

Being able to do these emergency medical procedures well will enable you to act as a first responder in dire circumstances. If someone is in shock or has a potentially fatal burn, your quick thinking and intervention can make all the difference. To increase your comfort and proficiency in providing emergency treatment, take the time to get familiar with these protocols and practice them frequently.

Medicinal plants for your natural first aid kit

You can easily obtain supplies for a natural first aid kit wherever you are—whether you're mowing your lawn or trekking up a mountain. Throughout the nation, medicinal herbs are widely available and have been utilized for ages to heal common illnesses and wounds.

"It's using our resources creatively and being linked to the land," Belfast, Maine-based clinical herbalist Steve Byers explained. "You can cure yourself without a Chinese-made Band-Aid."

Traditional first aid packs often include items like gauze, alcohol wipes, antihistamines, and athletic tape that are helpful in an emergency, according to him. However, they are not the exclusive choice. In the event of an emergency, such as being lost in the wilderness without your standard first aid bag, it might be useful to be aware of the resources around that could be used in the event of an accident or other medical emergency.

According to Byers, natural first aid for novices may be as easy as using burdock leaves instead of bandages to wrap an injury. or utilizing a tree's sticky, dried sap to remove splinters.

"Taking care of your health requirements, being able to respond to medical problems in the garden and the forest, it's important," said Greta de la Montagne, a Registered Professional Herbalist and Mentor with the American Herbalist Guild. "Putting medication back in the hands of the people is incredibly crucial."

De la Montagne, a holistic health practitioner in Montana, feels that knowing how to provide first aid naturally is powerful.

"We have been utilizing some of these therapeutic herbs for thousands of years," de la Montagne stated. "And more scientific research is beginning to demonstrate how these plants function now."

To help you get started, here are some plants:

Cattails

Similar to aloe vera gel, the jelly extracted from crushed cattail stems and leaves helps relieve burns, according to Byers. This widely recognized plant thrives across North America's wetland environments.

Furthermore, the plant's starchy roots are palatable and nourishing. Additionally, burnt and chaffed skin can be soothed and the danger of infection decreased by rubbing the mature "cobs" near the top of the plant.

Old Man's Beard

Bearded lichen, sometimes referred to as old man's beard, is a long, wispy strand that hangs down from tree branches and is traditionally used to treat bacterial infections. This lichen, scientifically known as Usnea, is found all throughout North America.

Saying of his native Maine, "It's the strongest antibacterial we have around here," Byers stated.

This lichen is frequently applied to wounds after being bathed in water. In order to be ingested or administered topically, it is also dried and powdered.

Usnea's therapeutic benefits have been shown in several investigations; one such research, conducted in 2004 at JSS College of Pharmacy in India, discovered the lichen to possess antiviral qualities. Ataturk University in Turkey conducted a research in 2004 that discovered antioxidant activity in an extract made from an elderly man's beard.

Jewelweed

According to a US Department of Agriculture plant profile, jewelweed is a tall plant with vivid orange blooms that is often found in damp, semi-shaded locations throughout northern and eastern North America. This plant, which can grow up to 5 feet tall, has long been used as a skin irritation cure.

When applied topically, the plant's sap from the stems and leaves is supposed to relieve stinging nettle and poison ivy-related discomfort and itching. It's occasionally used on hives and sores as well.

It's interesting to note that a 2012 study by Ohio State University researchers discovered that while jewelweed mash was successful in lowering poison ivy dermatitis, jewelweed extracts were ineffective. Thus, it seems that utilizing the plant fresh and in the field is the best course of action.

According to Byers, if you crush the stems for a few seconds, the sap will start to come out. Usually, this is put straight onto the skin.

Common plantain

The common plantain, which is native to Europe and certain regions of Asia, has spread throughout North America and is often thought of as an ugly weed. Nonetheless, has long been employed to calm skin and encourage wound healing. It's nourishing and edible as well.

This plant may be found anywhere—in most people's yards, on deserted lots, and along the side of pathways. I use it frequently to treat wounds. It's an excellent substitute for a Band-Aid.

In addition, she mashes it up to make a poultice, which she administers straight to wounds before covering them with leaves or a bandage.

Plantains, according to de la Montagne, "heal it up" when applied to wounds. When they return the next day, it will have nearly healed. It works so quickly that it's insane.

Plantain's ability to reduce inflammation was demonstrated in a 2015 study by researchers at the National University of Malaysia on rats, lending support to this application of the plant.

Plantains have also traditionally been used to treat digestive and respiratory problems. Actually, a 2003 research from Taiwan's Kaohsiung Medical University discovered that hot water extracts of the common plantain, or Plantago major as it is scientifically called, had a wide range of biological properties that would strengthen the immune system in addition to antiviral and anticancer properties.

Common yarrow

According to a botanical profile released by the USDA, common yarrow is a plant that grows throughout North America and is known for its clusters of little white flowers and feathery leaves. Native Americans utilized this plant to treat a wide range of diseases. This plant was crushed and put directly to burns and wounds; its leaves were also used to make a tea that was used to cure headaches and colds.

Achilles, who utilized plant extracts to heal his warriors' wounds during the Battle of Troy, is the inspiration behind the genus name Achillea of plants.

According to Byers, "it pretty much stops everything that's bleeding for minor cuts and abrasions quickly."

For this reason, Byers keeps supercharged yarrow in his first aid bag. Additionally, he has ground up the herb to make a "spit poultice" that he applies straight to wounds. He suggested that as an alternative, you could just mash the leaves with your hands.

Many studies have been done to look into the potential medical benefits of common yarrow. One such study, carried out in 2002 by experts in Greece and Turkey, revealed that an extract from the plant possesses antibacterial and antioxidant qualities.

Calendula

Calendula is not native to North America, although gardeners grow it all throughout the continent. This annual plant has fragrant, eye-catching orange and yellow blossoms that draw bees. It has also historically been utilized in complementary medicine to treat a variety of illnesses and wounds.

Based on a scientific assessment conducted by experts at Panjab University in India, Calendula officinalis is the only species of the genus Calendula that is widely utilized in therapeutic settings worldwide. It's frequently used to lessen inflammation and promote wound healing. It relieves rashes, burns, dry skin, and bee stings.

According to de la Montagne, "calendula should be kept on hand by everyone for survival reasons." The best way to use calendula is as a tea. I wipe my wounds with it.

De la Montagne advises individuals to gather calendula flower petals in the afternoon for personal use. After that, they can be dried for later use or steeped to make a tea or wash.

Arnica

According to an instructional website on the plant published by Penn State Health Milton S. Hershey Medical Center, arnica has long been used to relieve muscular pains, decrease inflammation, and heal wounds. The plant bears tiny, yellow, daisy-like blooms. Additionally, bruises and sprains are treated with it.

Arnica is a plant that is native to the mountains of Europe and Siberia and is grown all throughout North America. It is frequently used topically as a medication. If ingested, it can have detrimental negative effects. It's frequently available in tinctures, ointments, and creams.

Lavender

Lavender, a native of the Mediterranean region, is a popular herb in many North American gardens. This plant, which has lovely purple blooms and a pleasant, calming aroma, is frequently added to teas and made into extracts.

The National Center for Complementary and Integrative Health states that lavender has a long history of usage in medicine to treat anxiety and gastrointestinal issues. It has also been used to improve mood and appetite.

Furthermore, rats treated with lavender oil for wounds recovered much quicker from untreated wounds in a 2016 Japanese research. According to the researchers, "lavender oil has the potential to improve wound healing" as a result of this.

De la Montagne declared, "I adore using the lavender buds in a wound wash."

De la Montagne advises being cautious while using lavender, though, as some individuals may be sensitive to it.

White willow

The white willow tree, which is native to portions of Europe, Africa, and Asia, has been brought to much of North America. Its bark has long been used as a remedy for pain relief, particularly headaches, and inflammation reduction.

Willow bark has been shown in several in vitro and animal experiments to have anti-inflammatory properties. After salicin, the primary pharmacologically active component of willow bark, was isolated and refined in 1838, aspirin was finally discovered.

The most popular usage for the bark is in tea. But before using any medication, as with any other, make sure to speak with your physician. Allergies and upset stomach are a couple of willow bark adverse effects.

Conclusion

A first aid package that you are familiar with using is the best. It's critical to comprehend what's within and which things are multipurpose. It's also critical to understand the limitations of your first aid gear and how to use available resources in your surroundings. For example, I never bring a Sam Splint on a backpacking trip because I know there are plenty of creative ways to immobilize a joint using natural resources like trees and branches. With this knowledge, you can put together a first aid kit that is not only fully stocked but also customized to your particular needs and circumstances.

Keep in mind that the quality of your emergency treatment depends on the resources you have available. Maintain and refresh your first aid kit on a regular basis to ensure that it is prepared for any obstacles that may occur. Now that you have the knowledge and tools from this book, you are ready to be a dependable first responder and make a difference when it counts most. Remain vigilant, keep educated, and be prepared for any situation that may arise.

BONUS

101 Wilderness First Aid Tips and Tricks

Book 5

Solar Power Unplugged:
Off-Grid Energy Generation

Are you able to run your electronics off of the grid? Simple explanations of the fundamentals are provided in this tutorial, along with information on what you can (and cannot) realistically do and the equipment required to generate your own power for particular uses, such as charging a phone or flashlight without a wall outlet following a natural catastrophe.

It might be difficult to figure out how to keep your equipment operating without the grid. It's challenging enough to learn the jargon, but even if you grasp the fundamentals, you still need to put together a system that fits your budget and unique requirements.

The good news is that a graduate degree is not required. A lay of the land and a few basic terminology will guide your power readiness.

What's an Off-Grid Power System?

An off-grid power system transforms renewable energy sources, such as solar power, into electrical power that may be used. Since the off-grid system is independent of the local utility, blackouts won't have an impact on it.

Although solar power is a common option for off-grid systems, you may also utilize fossil fuels or other sustainable energy sources. The idea is that you don't need the grid to function.

In addition to solar generators, there are several more kinds available. In the end, you'll need to select the off-grid system and fuel source that best suit your requirements.

Differences Between Off-Grid & On-Grid Power Systems

Off-grid systems function independently of your local utility provider and are not connected to the utility grid. They need a power station since they are off the grid. It stores solar energy in the form of a massive battery by converting the DC power that the sun captures into AC (household) electricity.

Because on-grid solar power systems are connected to the utility grid, batteries are not always needed to store energy in them. When the system isn't generating enough power to fulfill your demands, they use the utility grid to get electricity instead of batteries.

Let's examine more closely at the key distinctions between the two configurations.

Access to Electricity

Since an off-grid power source is not connected to the electrical grid, it can only draw power when it produces its own energy or receives power from an other source. Solar panels are the source of energy generation for a solar-powered system since they transform sunlight into electrical power.

Battery storage is required for solar off-grid systems in order to store the extra power generated. During overcast days and at night, the system uses the energy stored in the battery as its main power source. In addition, EcoFlow solar generators have the ability to store energy from EV charging stations, auto adapters, and AC outlets.

With the exception of blackouts, on-grid systems are constantly connected to your local utility, so you will always have access to energy. Even if your solar system isn't producing enough energy to suit your demands, you may still enjoy power.

Excess Electricity Production Management

Batteries are typically used as a means of storing extra electricity for off-grid installations. Your house is powered by the electricity that has been reserved, even on cloudy or gloomy days.

In most cases, extra power generated by on-grid systems is returned to the grid. Owners of on-grid systems with net meters can get credits for the extra power they generate. You may reduce the cost of your power account by using these credits to pay for utilized electricity. On-grid systems without a net meter, however, frequently imply that you won't be able to use the extra power produced later.

Billing

After the setup is completed and paid for, off-grid solar systems operate on "free electricity." In the long run, they are more affordable and robust than fossil fuel-powered generators, even if they need a larger initial investment to pay for the required equipment. If you have to use gasoline or diesel to produce all of your power, the cost might fluctuate significantly and you have to refill frequently.

Utility services used by on-grid systems are paid for by the user. The cost of connecting to the grid is usually set, and there are extra charges for the power you consume after deducting any net meter credits.

Grid Power Outages

Off-grid systems are impervious to power interruptions that occur on the grid. Off-grid systems operate independently, so power interruptions won't interfere with your capacity to produce electricity. When using off-grid alternatives, users can continue to have power during prolonged blackouts as long as they are able to generate enough energy to meet their demands.

Unless you have a home backup power system in place, any power outage may cause disruptions in your access to energy for on-grid systems, which are dependent on the grid's utility.

Benefits of Using an Off-Grid Power System

Because off-grid power systems have so many advantages, they are a desirable option for campers, RVers, and anyone pursuing energy independence.

Energy Self-Sufficiency

With off-grid power solutions, you may live anywhere there are enough resources to charge your devices. It frees you from being a slave to the local electricity utility and allows you to live and work in isolated places. When the town is devastated by a power outage, you are unaffected.

Diminished Carbon Emissions

Systems for off-grid solar electricity enable you to have an eco-friendly lifestyle. Compared to on-grid and fossil fuel generators, several off-grid energy sources, such solar, water, and wind, are far more environmentally benign. Your carbon footprint is lowered since these renewable energy sources don't rely on fossil fuels.

Reduced Energy Prices

Consider the possibility of living without electricity bills. You will be dependent on nearly endless, pure natural resources for energy generation if you live an off-grid renewable energy lifestyle. You won't need to use the grid if you have battery storage. You don't have to worry about skyrocketing power costs if you have enough ability to produce enough electricity to live off the grid.

Installing Quickly

An off-grid solar system consists of comparatively basic parts. On the other hand, grid-tied systems require labor expenditures to employ specialists for installation due to their complexity.

Solar power setup and maintenance

While the cost of installing an off-grid solar system is continually growing, the cost of solar panels has been gradually declining. But anybody with a toolkit and a rudimentary understanding of electricity can install it themselves. You will learn a lot and the overall system cost will be significantly decreased.

To assemble a simple off-grid solar system, you will require the accompanying parts:
- Solar panel
- Charge Controller
- Battery
- Inverter
- Balance Of System (Cable, Breaker, Meter, Fuses, and MC4 connectors)

Supplies
- Solar Panel
- Charge Controller

- Battery
- Inverter
- Remote Meter
- WiFi Adapter
- Temperature Sensor
- DC Breaker
- AC Breaker
- DC Busbar
- Fuse Box
- DIN Rail
- Cables
- MC4 Connector
- Terminal Lugs
- Cable Tie

Tools Required :

- Wire Stripper
- Crimping Tool
- Plier
- Screwdriver
- MC4 Spanner
- Spanners

How It Works ?

With an off-grid solar system, you are completely cut off from the utility grid. Batteries are used by the system to store solar-generated energy.

Solar Panel: The solar panel generates energy from solar radiation. The solar panel's photovoltaic cells capture solar energy and transform it into DC power.

Charge Controller: A charge controller regulates the amount of current that flows to a battery using the electricity that comes from the solar panel. Batteries cannot be overcharged or overdischarged thanks to charge controllers.

Battery: During the day, it stores the energy produced by the solar panel.

An inverter is a device that changes DC (Direct Current) electricity from solar panels or a battery bank into AC (Alternating Current) power so that AC appliances, such a refrigerator, TV, fan, and water pump, may be used.

Basic Electricity Rules

Ohms Law Relationship

Current (I) = Voltage (V) / Resistance (R)

The Ohms law connection is simpler to recall when you use the Ohms output above. By using Ohm's Law, we may determine the third missing value if we know any two values of the voltage, current, or resistance variables.

Voltage (V) :

V (volts) = I (amps) x R (Ω)

Current (I) :

I (amps) = V (volts) / R (Ω)

Resistance (R) :

R (Ω) = V (volts) / I (amps)

Power (Watt) = Voltage (Volt) x Current (Amp)

Energy (Watt-hours) = Power (Watts) $\times$ Time (Hours)

Capacity = Current (Amp) x Time (Hours)

DIY Off-Grid Solar

The following steps are required for building a DIY Off-Grid Solar System:

Calculate Daily Energy Consumption

Embarking on the off-grid solar journey starts with a simple equation:

Energy Consumption (Watt-Hours) = Power (Watts) $\times$ Time (Hours)

Now, where to find that power rating? Peek at the appliance's power label or nameplate, or measure it with a trusty wattmeter. I've personally used my wattmeter to measure various appliances.

Manual Calculation Scenario:

Let's do the math for running:

2 LED bulbs (6W each) for 5 hours a day

1 Fan (80W) for 4 hours

1 Laptop (65W) for 3 hours

1 WiFi Router (6W) for a continuous 24 hours

1. LED Bulb: 2 x 6W x 5 hr = 60WH

2. LED TV: 1 x 65W x 3 hr = 195WH

3. Ceiling Fan: 1 x 80W x 4 hr = 320WH

4. WiFi Router: 1 x 6W x 24 hr = 144WH

Total = 719WH

Select the Battery

In the realm of off-grid solar systems, the battery takes the spotlight—capturing and storing the energy produced by the solar panel during the day. It's not just a component; it's the lifeline that ensures a steady stream of reliable power when the sun bids adieu.

However, choosing the right battery is no small feat, especially considering its significant contribution to the overall project cost. Let's delve into the details to guide you in selecting the perfect battery for your off-grid solar setup.

Batteries, in this context, fall into two key categories:

1. Applications: Automotive and Deep-Cycle

2. Chemistry: Lead Acid, Lithium, and NiCd

Automotive Battery:

This kind of battery is intended to deliver a significant quantity of electricity for a brief length of time. In order to start the engine, this current spike is necessary. In order to maximize surface area and, thus, higher beginning current in starting batteries, several thin plates are used.

Application: Automobiles (Car & Bike)

Deep-Cycle Battery:

A deep cycle battery is made to deliver a constant current for an extended length of time. This kind of battery is also intended to be repeatedly thoroughly drained. A deep cycle battery employs larger plates to do this. Unlike the beginning batteries, this will result in lower surfaces and hence less instantaneous power.

Application: Renewable Energy

Lead-Acid Battery Vs Lithium-Ion Battery

Lead-acid and lithium-ion batteries are two of the most used battery chemistries. NiCd is also utilized for renewable applications in addition to these; however, I will only cover the first two in this discussion.

Lead is used to make lead-acid batteries, whereas lithium is used to make lithium batteries. Lead-acid and lithium batteries are both capable of efficiently storing energy, but they each have certain benefits and limitations.

1. Lead-acid Battery:

An established technology that is less expensive but requires frequent maintenance and has a shorter lifespan is the lead-acid battery.

Lead-acid batteries classified as flooded (FLA) are immersed in water. To keep them functioning correctly, they need to be replenished every one to three months and examined often. To allow battery gasses to escape, it must also be positioned in a ventilated area.

Sealed Lead-Acid (SLA) Batteries: AGM (Absorbent Glass Mat) and Gel are the two varieties of SLA batteries, and they share a lot of characteristics. They are spill-proof and require little to no upkeep. The main distinction between gel and AGM batteries is that the former typically have lower power and charge rates. Gel batteries often have a lower capacity and a slower rate of recharging due to their inability to withstand high charge currents.

2. Lithium Battery:

Although lithium is a more expensive battery technology due to its longer lifespan and higher efficiency, the performance gain is worth it.

Lithium Iron Phosphate (LiFePO4) batteries are used in solar systems because of their excellent thermal stability, high current ratings, and extended lifespan. This new technology can be used in deeper cycles and has a longer lifespan. Moreover, they don't need to be vented or maintained like lead-acid batteries do. The primary drawback of lithium batteries at this time is their current price premium over lead-acid batteries.

Which Battery Should You Choose?

Lead-acid and lithium-ion batteries are both viable choices if you want a battery backup solution. Nonetheless, considering the many benefits of lithium-ion technology, such as increased efficiency, longer lifespan, and better energy density, installing one is typically the best choice.

Choose Flooded Lead Acid (if you don't mind frequent maintenance) or the premium Lithium option for heavy use if you intend to live off the grid permanently.

You won't spend a lot of time there if you install solar in a vacation house or tiny cottage. In this situation, you won't be able to give the flooded lead-acid batteries the routine maintenance they need. Then, I would advise shelling out a little more money to purchase a sealed lead acid battery.

Factors Determine the Battery Bank Size

The following factors determine the battery bank size:
1. Daily power consumption
2. System voltage (12V / 24V /48V)
3. Depth of Discharge (DOD)

System Voltage

In the world of batteries, understanding the volts (V) and amp-hours (AH) is the key to unlocking a reliable power source. To achieve the desired system voltage, you can employ a strategic dance of wiring—either in series or parallel.

Series Connection:

Think of series connection as a tandem ballet. It adds up the voltage, creating a harmonious duet, while the amp-hour rating (or Amp-Hours) stays consistent. For instance, intertwining two

12V /100AH batteries in series creates a powerful 24V performance, yet the total capacity remains a steadfast 100AH.

Parallel Connection:

Parallel connections are the ensemble cast, boosting the current rating (Amp-Hours) without altering the voltage. But beware, with increased amperage comes the need for sturdier cables to prevent a cable meltdown. For instance, uniting two 12V /100AH batteries in parallel maintains a 12V rhythm, yet elevates the total capacity to a robust 200AH.

Depth of Discharge

The proportion of the battery's capacity that may be safely drained without causing damage to the battery is known as the Depth of Discharge, or DOD.

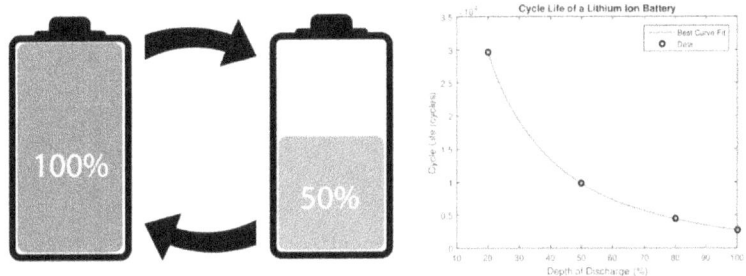

The above graphic illustrates how a battery's lifetime decreases with increasing drain. Although deep cycle batteries are intended to be discharged to 80% of their capacity, it is advised to select a value of about 50% as a fair compromise between cost and longevity.

It is generally accepted that a deep cycle battery should have 50% DOD, while a lithium battery should have 80% DOD.

Battery Sizing

Battery Capacity (AH) = Daily Energy Consumption (Watt-Hour) / (System Voltage x DOD)

For instance, if your daily energy consumption is 719 Watt-Hours (as calculated earlier), your system voltage is 12V, and you're working with a Depth of Discharge (DOD) of 50% for a Flooded Lead Acid Battery, the calculation goes like this:

Battery Capacity = 719WH / (12V x 0.5) = 119.8AH

Now, you can't buy a fractional battery, so rounding up, you'd look for a battery with a capacity of more than 119.8AH. In the market, the closest available value is 120AH. So, you've selected a battery:

Battery Selected: 12V / 120AH

But hold on! Thinking ahead, considering future expansion, you've opted for a bit more power, purchasing a 150AH battery.

With this, your off-grid solar system is armed and ready to illuminate the night, providing a reliable source of power for your needs.

Select the Solar Panel

Sunlight is converted into power by solar panels. Although solar panels are not 100% efficient and cannot capture all of the sun's energy, they can convert a certain portion of that energy into power. Less than 20% of solar panels are efficient, meaning that only 20% of the energy from the sun can be captured by them.

Commonly they are 3 types:

1. Monocrystalline:

Due to its single silicon source, monocrystalline solar cells have a higher efficiency.

Compared to other solar panel types, monocrystalline solar cells have a better efficiency because they are composed of a single silicon crystal, which facilitates easier electron passage across the cell. There is a range of 17% to 22% efficiency.

Monocrystalline solar panels are more expensive than other types of solar panels due to the manner they are made.

2. Polycrystalline:

Blended from many sources of silicon, polycrystalline solar cells have a little lower efficiency. Electron flow is impeded in each solar cell by the numerous silicon crystals. Polycrystalline panels have a lower efficiency rate than monocrystalline panels due to their crystal structure. Efficiency ratings for polycrystalline panels usually lie between 15% and 17%.

The cost of producing polycrystalline solar panels is lower than that of monocrystalline panels. Polycrystalline solar panels are used in the majority of home installations.

3. Thin Film:

A thin coating of a photovoltaic material is deposited onto a solid surface, such as glass, to create thin-film solar panels. Amorphous silicon (a-Si), cadmium telluride (CdTe), copper indium gallium selenide (CIGS), and dye-sensitized solar cells (DSC) are a few examples of these photovoltaic materials.

Amorphous solar cells' primary benefit is its ability to produce energy in low light. But the primary issue with amorphous solar cells is their meager photoelectric conversion efficiency, which is only between 10 and 13 percent.

Which One Should You Select?

Installing monocrystalline solar panels makes the most sense for the majority of household solar panel installations. Compared to polycrystalline panels, you get superior efficiency and a sleeker appearance, but at a higher cost.

However, polycrystalline panels can be a better option if you're on a restricted budget.

Due to their lower efficiency ratings, thin-film solar cells are mostly employed in large-scale activities like utility or industrial solar systems.

I will consistently advise getting a solar panel from a reputable brand. A reputable solar panel manufacturer will always make significant investments in both its reputation and the caliber of its production process.

Factors Determine the Solar Panel Sizing

The sizing of the solar panel used in an off-grid system depends on the following factors:

1. Daily energy consumption
2. Number of Peak sun hours
3. Solar panels efficiency

Peak Sun Hours

Finding out how much sunshine your home receives is the first step in choosing the appropriate size for the solar panel. Peak sun hours are a more accurate indicator of the quantity of energy solar panels can generate, even though the amount of sunlight they get is still crucial.

Peak Sun-Hours: What Is It?

The number of hours per day when the average solar irradiance, or sunshine, reaches 1000 watts per square meter (W/m2) or 1 kilowatt per square meter (kW/m2) is known as the peak sun hours.

One peak solar hour is equal to 1000 W/m2 or 1kWh/m2.

For instance, a site obtains 6.65 peak sun hours if it receives 6,650 Wh/m2 of solar energy on a particular day. You can clearly comprehend this by looking at the photo above.

The quantity of peak solar hours is influenced by the following factors:

1. Geographical Location: The quantity of sunshine received by solar panels deployed at different locations varies. Because it is located closest to the sun, the panel placed toward the equator receives the most sunlight.

2. Time of Day: Depending on the sun's location in the sky, the solar panel receives different amounts of sunlight throughout the day. It gets the most during midday and the least in the morning and evening.

3. Season: Summertime maximum solar radiation and wintertime minimal radiation.

How to Calculate Total Peak Sun Hours?

In the world of solar energy, understanding how much sunlight your location basks in is key. Enter the solar irradiance map, a guide to the average solar energy your area receives on its gloomiest day.

To uncover this information, follow these steps using the Global Solar Atlas:

Step-1: Search your location.

Step-2: Choose the PV system configuration (e.g., Small residential).

Step-3: Click on Annual Average (Daily Average in kWh/m2 per day).

Step-4: The number revealed is the coveted peak sun hours.

As an example, let's take New Delhi, India, boasting 5.093 kWh/m2 per day according to the Global Solar Atlas. Now, considering solar panels are rated at an input of 1 kW/m2, we can calculate the peak sun hours:

Peak Solar Radiation = 1 kW/m2

Peak Sun Hours = 5.093/1 = 5.093 Hours

In the spirit of preparedness, let's be a bit conservative. For worst-case scenarios, choosing a number slightly lower, say 4.5 hours, ensures your solar setup can handle even cloudy days.

For those in North America, a handy reference chart is available to estimate the number of peak sun hours.

How to Wire Solar Panels?

There are three methods that you can connect the solar panels:

1. Series: The panels' voltages are added to one another while the current stays constant.
2. Parallel: The number of currents is added while the voltage stays constant.
3. Series-Parallel: Parallel connections between strings of series panels

Two crucial factors to take into account while choosing amongst the aforementioned three configurations are the charge controller's maximum input voltage and type (PWM or MPPT).

MC4 connectors and spanners are required to connect and disengage the solar panels in the three combinations mentioned above.

Select Charge Controller

It is a device that regulates the quantity of electric energy generated by solar panels that enters the batteries. It is positioned between the solar panel and the battery bank. Making ensuring the battery is correctly charged and shielded from overcharging is the primary purpose.

The charge controller controls the battery charge to prevent overcharging and disconnects the load when the battery is fully depleted as the input voltage from the solar panel increases.

Types of solar charge controllers

There are currently two types of charge controllers commonly used in PV power systems :

1. Pulse Width Modulation (PWM) Controller
2. Maximum Power Point Tracking (MPPT) Controller

PWM Solar Charge Controller

Pulse Width Modulation, or PWM for short, refers to the technique it utilizes to control charge. Its purpose is to guarantee that the battery is fully charged by lowering the solar array's voltage to almost that of the battery. Stated differently, the solar panel voltage is locked to the battery voltage by pulling the solar panel voltage sensor down to the battery's system voltage while maintaining the same current.

The solar panel and batteries are connected and disconnected by an electronics switch called a MOSFET. The MOSFET may be switched at a high frequency with different pulse lengths to maintain a steady voltage. By altering the widths (lengths) and frequencies of the pulses delivered to the battery, the PWM controller self-adjusts.

The MOSFET is fully ON when the width is 100%, which enables the solar panel to bulk charge the battery. After the battery is fully charged, the transistor is turned off when the width is at 0%, open-circuiting the solar panel and stopping any current from reaching the battery.

MPPT Solar Charge Controller

The Maximum Power Point (MPP) is the voltage at which the PV module must run in order for the MPPT charge controller to harvest the maximum power from the PV module. Its design allows it to adapt the maximum power output of the solar array to meet changing voltage requirements by adjusting its input voltage. A DC/DC converter is used to change the input voltage.

In order to maintain the most efficient power output for the system, MPPT controllers employ an adaptive algorithm that tracks the maximum power point of the solar panel or array and then modifies the incoming voltage.

At low temperatures (below 45°C), an MPPT controller offers a significant performance improvement (10% to 40%). Compared to the PWM controller, they are more efficient. An MPPT controller typically has an efficiency of 94–99%.

The array voltage needs to be significantly greater than the battery voltage in order to utilize the MPPT controller to its fullest capacity. For greater power systems, the MPPT controller is the ideal choice.

Sizing of Charge Controller

Choosing the most suitable charge controller requires two steps:

1. Voltage Selection: The voltage of the charge controller and the system must coincide. 12, 24, and 48 volt versions are the typical ones. A charge controller with a 12 volt rating is required if you are wiring your batteries at that voltage.

Certain controllers are voltage specific, which means that the voltage is fixed and cannot be adjusted. Some more advanced controllers may be utilized with various voltage settings thanks to a function called voltage auto-detect.

2. Current Selection: You must be aware of the battery voltage and the solar panel's maximum output current in order to choose the appropriate charge controller.

The system's maximum allowable current is equal to (system voltage / solar panel wattage) × safety factor.

Safety Factor: To account for all solar panel output-boosting conditions, such as a bright day with a very clear snowpack, we employ a standard factor. (more light bouncing off the snow). That amounts to 1.3, or 130%.

PV. Charge Controller: Max Input Voltage

There is an upper voltage restriction on charge controllers. This is the highest voltage from the solar array that they are able to withstand. It's important to be aware of the maximum voltage limit and to avoid beyond it to avoid damaging your solar charge controller.

Sample Calculation

Sizing Up Solar Control: A Guide to Charge Controller Selection

When it comes to charging your 12V battery bank with a 260W solar panel, the key lies in choosing the right charge controller. Let's break it down step by step.

1. Voltage Rating:

Match the charge controller's voltage rating to the system voltage. In this case, it's a 12V battery bank, so the controller's voltage rating should be 12V.

2. Current Rating:

Now, let's determine the controller's current rating, factoring in a safety margin (safety factor = 1.3). The formula goes like this:

Rating = (Solar panel Wattage / System voltage) × Safety factor (1.3)

Rating = (260W / 12V) x 1.3 = 28.16A

Now, you can't have a fraction of an amp for a controller, so rounding up, the solar charge controller rating is selected as:

Solar Charge Controller Rating = 30 Amps / 12 Volt

With this careful selection, your solar charge controller is primed to efficiently manage the flow of power from the solar panel to your 12V battery bank.

Select Inverter

One of the most crucial parts of a solar panel system is the solar inverter. They are in charge of transforming the alternating current (AC) from your solar panels' direct current (DC) electricity into power for your appliances.

You can omit this step if you are solely using your battery bank to power DC loads. However, you must change the direct current from the batteries into alternating current for your appliances if you are powering any AC loads.

Common Types Of Inverter:

1. Square Wave
2. Modified Sine Wave
3. Pure Sine Wave

Although the square wave inverter is the least expensive, not every device can use it. Additionally, several appliances, including microwaves, refrigerators, laser printers, sensitive electronics, and most types of motors, are not compatible with modified sine wave output.

Compared to pure sine wave inverters, modified sine wave inverters often operate less efficiently.

Thus, in my view, go with a pure sine wave inverter.

Selecting the Solar Cable

Efficient energy transfer from solar panels to the battery is paramount, and this involves understanding and mitigating cable resistance. According to Ohm's law, the voltage drop (V) due to cable resistance (R) is given by $V = I \times R$, where I is the current.

The cable's resistance (R) hinges on three crucial factors:

Cable Length: The longer the cable, the higher the resistance.

Cable Cross-section Area: A larger area translates to lower resistance.

Material: Copper or Aluminum. In this scenario, copper, with its lower resistance, is the preferred choice.

To tailor this to your needs, provide the following parameters:

Solar Panel Operating Voltage (Vmp)

Solar Panel Operating Current (Imp)

Cable Length from Solar Panel to Battery

The expected loss in percentage

The specification sheet located on the rear of the solar panel or the datasheet are a good place to start looking for the first two parameters, Vmp and Imp. The installation determines the length of the cable. A good design is defined as a loss percentage of between two and three percent.

We have already decided on the solar panel and the rating in the previous phase. Vmp = 36.7V and Imp = 6.94A (rounded to the nearest higher figure, i.e., 37V and 7A) are from the solar panel specification sheet. Assume there is a 30-foot gap between the solar panel and the battery, with a 2% anticipated loss. The cable size is 12 AWG when the RENOGY online calculator is used with the aforementioned settings.

How to Maintain an Off-Grid Solar System

An off-grid solar system is, as the name implies, one that is not wired into a utility grid. Through solar panels that store energy in a battery bank, it is able to produce power.

Caring for the battery bank is the most crucial aspect of off-grid solar system maintenance. This can lower the overall cost of your RE system and increase the lifespan of your batteries.

Check the charge level.

The amount that a battery has been depleted is indicated by the depth of discharge (DOD). The exact reverse is true for the state of charge (SOC). In case the DOD is 20%, the SOC would be 80%.

Don't let the battery discharge any farther than this because doing so frequently can reduce its lifespan by more than 50%. To find the battery's SOC and DOD, measure the voltage and specific gravity.

An amp-hour meter can be used for this purpose. Nonetheless, a hydrometer provides the most precise means of determining the fluid's specific gravity within.

Equalize your batteries.

A battery bank is made up of numerous batteries, each containing several cells. The specific gravity of the various batteries may change after charging. One method of maintaining full cell charge is equalization. It's common advice from manufacturers to equalize your batteries once every six months.

You may configure the charge controller to perform equalization on a periodic basis if you would prefer not to continually check your battery bank.

You might be able to choose the equalization process's duration and a certain voltage using the charger.

To ascertain whether your battery bank requires equalization, you may also do it manually. Using a hydrometer, determine each cell's specific gravity and note whether any are noticeably lower than the rest. If so, make sure your batteries are balanced.

Check the fluid level.

Sulfuric acid and water are combined to create flooded lead-acid (FLA) batteries. A portion of the water evaporates when the battery powers up or charges. When utilizing a non-sealed battery, you must top it out with pure water; however, sealed batteries do not have this issue.

Take off your battery cap and see how much fluid is there. Add distilled water until the surfaces of the metal lead disappear. To prevent spills and overflow, most batteries require a fill guide.

Each cell should have a hydrocap installed in lieu of the old cap to stop water from escaping too rapidly.

To keep dirt from getting into the cells, make sure the battery's top is clean before removing the cover.

Battery consumption will determine how frequently you top off. There may be greater water loss with heavy charging and loading. Once a week, check the fluid for fresh batteries. You may determine how frequently to add water from there.

Clean the batteries.

Some water may form condensation on top of the battery as it exits through the cap. Because of its modest acidity and electrical conductivity, this fluid can draw more current than is necessary and form a little channel between battery posts.

Use distilled water and baking soda mixed with a specific brush to clean battery connections. Make sure all connections are secure and give the terminals a quick rinse with water. Apply a high-temperature grease or commercial sealer to the metal components. Take care to prevent baking soda from entering the cells.

Do not mix batteries.

Replace a whole quantity of batteries at a once. Combining new and old batteries can lower performance since the fresh ones soon lose quality and become equal to the older ones.

Your off-grid solar system's longevity and efficiency may both be increased with proper battery bank maintenance.

Conclusion

There are many practical use cases for an off-grid power solution, such as camping, living in your RV or tiny home, or simply home backup power storage. Off-grid solar systems give you the energy independence you need to live off-grid.

As we wrap up our journey through "Solar Power Unplugged: Off-Grid Energy Generation," we're reaching the end of our adventure into a world where sunlight becomes our own personal power plant. From the start, we believed that the sun could be our energy hero, freeing us from the usual power setups. Now, as we finish, the dream of a world where we're not tied to traditional energy sources is clearer.

Throughout our time together, we've uncovered how solar power works, making the complex technology easy to understand. We've learned how it can change the way we create and use energy, putting the control back into the hands of everyday people and communities. It's not just about using solar panels; it's about storing that energy in advanced batteries, creating a new way of life where we're not dependent on regular power companies.

Living off the grid isn't just an idea; it's a real, doable thing. We've explored the stories of folks who have chosen to live without relying on the usual power systems, creating homes where solar power isn't just a choice but the main force behind a life of independence and strength.

As we say goodbye, let's remember that solar power isn't just a technology; it's a force for change. It gives us the ability to shape our energy future, promoting independence, eco-friendliness, and a deep connection with the world we live in.

This isn't the end of the road; it's a new beginning. The sun will keep shining, and our ability to capture its energy will only get better. So, whether you're a seasoned off-grid pro or just starting your solar journey, know that the possibilities are as endless as sunlight itself.

May your path be brightened by self-generated power, and may your journey inspire others to see that the future of energy is unplugged.

Cheers to a cleaner, brighter, and more independent future.

Unplug, capture, thrive.

BONUS

DIY Off-Grid Solar FULL Install & Wire Diagrams

Book 6

The Modern Homesteader's Handbook: Mastering Self-Reliance Skills

Have you heard the term "modern homesteading" spoken around recently but you are unsure of its definition? With this book guide, you are going to know what it really means to be a modern-day homesteader.

Modern Homesteading

Modern homesteading is having a huge moment right now, as you may have noticed if you've been browsing Instagram recently. Homesteading is becoming a popular option for people and families looking to live a more deliberate, straightforward lifestyle. People want to know precisely where their food comes from, whether they are opting to buy from local farmers, create a home farm, or relocate to the country. Families are becoming more interested in homesteading as a result of a greater emphasis on hard labor, health, and having a relationship with their food.

However, what exactly is the "modern homesteading" movement all about? Although the solution is a little more nuanced than a straightforward explanation, let's get started.

What is "modern homesteading," then? Well, maybe homesteading—a more generic term—should be our first choice. For many years, the phrase denoted both the abilities required for pioneer life and a free government land program. Homesteading is now more often used to describe a way of life that encourages increased self-sufficiency. A little ambiguous, yes? It is evident that the phrase "homesteading" lacks a precise meaning. Instead, a person may design their own homestead according to what works best for their family. Rather than being based on a rigid set of rules, homesteading is really a lifestyle choice.

Putting the goal and aim of our own farm front and center may be the simplest way to sum up the current homesteading movement. Homesteading is defined as making the most of one's abilities to live off the land. There are a plethora of approaches one may take when it comes to homesteading. It might be as easy as gathering eggs from backyard hens or as difficult as operating a fully functional farm and living off the grid. The phrase "homesteading" is more of a continuum than an absolute. It's a manner of living that centers on one's goal and purpose, which may be anything from feeling more connected to the processes of food production and growth to living a more sustainable existence.

Homesteading basics

A list of supplies for first-time homesteaders

Do you know what supplies you'll need to begin a homestead? In this part, you'll get a list of supplies for homesteading. Ideal for first-time homesteaders!

It may be really intimidating to start a homestead from scratch, particularly if you've never done it before. This collection of tools for unfamiliar homesteaders will help you with anything from basic setup to emergency preparedness.

Food supplies

One of the most crucial things on my list of homesteading supplies is definitely food. Purchasing the following products is a smart choice if you're just starting out as a homesteader:

Metal Storage Container: Use metal containers to store food to keep rats away.

Feed Scoopers: There's always room for more scoopers! These scoopers are essential for scooping pellets, grain, black oil sunflower seeds, and other foods.

Extra Feeders: Like scoopers, feeders are something you can never have too many of. They work well for hanging pellet feeders on your metal fence or for holding loose minerals. The options are endless!

Buckets: One more time, you can never have too many buckets on hand.

Heated Waterer: A heated waterer is a must if you live in a cold environment and want to keep the water from freezing over for your animals. This one is for goats, and this one is for hens.

Housing Supplies

Building a homestead from scratch may be a huge undertaking for first-timers. The following goods should always be accessible homesteading supplies:

Heat Lamps: Tried to avoid utilizing heat lamps as much as possible since they provide a fire risk. That being said, if you have a few brooder boxes full of ducklings and chicks in the barn, you may sometimes need a heat light. This one is used to lower the danger of fire.

Fencing: Whichever fence you decide on, be sure it is suitable for your predator scenario and safe for your specific animals, electric poultry netting is mostly used.

Automatic Coop Door: This is a significant investment, but well worth it. You don't have to bother about releasing your hens outside every morning and putting them away every night if you set the coop door to open or shut every morning or evening. Major time saver!

First Aid Kit

You most likely don't know what should be in a first aid kit if you're a novice homesteader. While by no means comprehensive, make sure to have these supplies on hand whenever you go homesteading since you can never be too prepared.

- Spray Bottle
- Food Grade DE
- Syringe
- Thermometer
- Scalpel (useful to have around in case your chicken gets bumble foot)
- Bandage Tape
- Bandage Wrap
- Corid
- Poultry Drench
- Toltrazuril

Ensure that you have the following things on your homesteading supplies list on hand in case your goats decide to kidding:

- Oral Drencher
- Towels
- Gloves
- Bulb Syringe
- Hair Dryer
- Bottles
- Pritchard Nipples
- Battery Powered Lanterns
- Iodine
- Calcium Drench

Supplements

Your cattle may be more susceptible to mineral and other dietary deficits depending on the kind of animal you grow. If you're a first-time homesteader who intends to raise goats and/or ducks, you may want to consider adding the following supplies to your list.

Niacin (Ducks)

Replamin Gel Plus + Gel/Paste Applicator (Goats)

Copper Bolus (Adult goats here; kid goats here)

How To Start A Homestead

It's usually a gradual transition to become a homesteader from a standard contemporary lifestyle. It's not necessary to sell everything and go to the country right away.

Many individuals have romanticized, idealized ideas of what it would be like to live on a farm.

Step 1: Consider What Homesteading Involves

It would be wise for you to take a moment to consider the daily tasks and responsibilities that come with choosing to live on a farm.

Particularly tending to crops and animals requires a lot of time and physical effort, therefore not everyone is suited for it.

It's important to confirm if homesteading is the sort of life you both want to lead if you have a spouse or other significant other.

It will be necessary for you to sit down and discuss your goals in an honest and transparent manner. Your companion will find it quite challenging to live a homestead lifestyle if they detest being hands-dirty.

Before making any kind of commitment, you should invest many hours in studying all you can about homesteading.

Don't make a significant homesteading choice until you have all the information you need. Read books, watch films, and become completely absorbed in the homestead way of thinking.

Try to offer to assist out for a few days if you have friends or relatives who already own a farm so they can give you an idea of what it's like. Additionally, make sure you grill them extensively.

Step 2: Set Goals For Yourself

The amount of the property you need will depend on your objectives from Step 2. If you intend to continue working a full-time or part-time job and homestead just as a hobby, you can most likely survive in an urban or semi-rural setting.

You'll need enough land to produce all the fruit and vegetables you want, as well as room for cows, lambs, or any other animals you like, if you intend to make homesteading your full-time occupation and way of life.

You should determine the approximate region in which you want to reside in addition to the amount of land you really require.

- Would you want to cut your carbon impact down to a certain percentage?
- Would you like to live totally off the grid, partly off the grid, or on the grid?
- Do you wish to grow fruit trees, rear cattle, or do other activities that call for additional land?

You'll be better able to plan your next course of action once you have an idea of what you desire.

Step 3: Decide Where You Want To Live

The amount of the property you need will depend on your objectives from Step 2. If you intend to continue working a full-time or part-time job and homestead just as a hobby, you can most likely survive in an urban or semi-rural setting.

You'll need enough land to produce all the fruit and vegetables you want, as well as room for cows, lambs, or any other animals you like, if you intend to make homesteading your full-time occupation and way of life.

You should determine the approximate region in which you want to reside in addition to the amount of land you really require.

Do you want to live just outside of town, or are you alright living in a more isolated area?

Verify that the property you are considering will really support the kind of homestead lifestyle you want to lead.

For example, excessively sandy or rocky soil will make things more difficult if your main goal is to cultivate crops.

Remember to account for travel time. If you still want to have a career, do you really want to travel 1.5 hours each time you need to pick up groceries or go to work every day?

Do you accept that in an emergency, it can take an hour for the police or an ambulance to arrive?

Little things like having to drive to the closest post office once a week or take a lengthy walk to your mailbox every day might be more than you bargained for.

Additionally, resist the urge to bite off more than you can chew. To have the farmhouse of your dreams, you don't even need ten or even one hundred acres.

Usually, 2 to 5 acres will be more than sufficient to meet the needs of a single family. Anything larger and you could discover that maintaining it is just not worth the hassle.

Some important homestead factors to keep in mind during the planning stage include:

Water access. Are there any ponds, rivers, or lakes close by that you might utilize for water? Does the land have a well? What is the annual rainfall in the area?

Land safety. If you are producing your own food, you should avoid living in an area that has frequent droughts as well as being close to oil drilling operations or other possible health risks.

Community. Sometimes the property you purchase is not as essential as the community you live in. It will be necessary for you to network and establish friendships with locals. It might be more challenging to blend in if they have different political or religious beliefs than you, particularly in a small hamlet.

Shool. Is there a school in the area if you have children? If not, you could have to teach them at home.

Step 4: Make A Budget

When homesteading, having a well-planned budget is essential, especially if you want to leave a stable career to become fully self-sufficient.

Avoid using all of your funds when purchasing land or other real estate. If not, you won't have any money left over for equipment, upgrades, renovations, or other essentials.

Any repairs or enhancements to your property should, in general, be expected to cost 50% more than you anticipate and take twice as long.

You will need to come up with some ways to make money for yourself if you are quitting your work in favor of a more independent lifestyle.

At the very least, you will probably still have to pay utilities, property taxes, and other bills for things like phone or internet service.

Additionally, you should save up some money in case of an unforeseen circumstance, such a furnace breakdown or a family member being ill.

Having many sources of income from your farm is a wise move. You might attempt selling items like soap-making or other crafts, wool, milk products, and surplus food.

In this manner, you have backup plans in case your crops all perish or you discover that there isn't a market for one of your revenue streams.

It goes without saying that you don't want to overextend yourself. However, it's not unheard of for homesteaders to have five or ten distinct goods or revenue streams.

Step 5: Start Small

You may start now, without waiting to build your ideal farm. You are able to begin homesteading immediately. Homesteading is more about a way of life and mentality than it is about where you live.

Regardless of your circumstances, including apartment living, you may take the first steps this week toward living a more independent existence.

You may begin growing your own lettuce or herbs inside if you have a sunny window.

Do you have a sizable backyard where you only grow weeds and grass?

Next spring, install a raised bed or garden and begin producing some of the veggies for your family. (Make sure the veggies you choose are ones you love eating on a daily basis!)

Do you presently have a fireplace that you are not using? It's time to start utilizing wood to lower your heating costs by cleaning out your chimney and getting some!

You may progressively add more and more tasks over time. Over time, the little lifestyle adjustments you make each year will really start to add up.

In your backyard, you may even begin keeping bees or poultry. Just be sure to first confirm with your local bylaws that it is permitted!

Homesteading is all about doing what your gut tells you to do. You are free to decide what is most important to you and what sequence to complete tasks in.

Some individuals would prioritize becoming energy independent, therefore they might want to get solar panels right soon.

It may not bother other individuals to pay for their gas and electricity. For ethical considerations, some individuals may decide never to embark down the path of keeping cattle for the sake of producing eggs and meat, while others would desire to begin doing so immediately.

Continually Simplify Your Life

Being more frugal and minimalistic are typically associated with homesteading.

Breaking the loop of always wanting the newest and best phones, gadgets, fashionable clothes, and other items that might drain your bank account but don't actually give much value is a major part of that.

Less is more for homesteaders, because there's generally a better, less expensive method to do a task.

Regularly doing an audit of your life will help you identify the items that are consuming a lot of your money, time, and energy and determine if you can cut them down or get rid of them entirely.

It's usually necessary to get go of certain items before incorporating homesteading into your lifestyle. Certain things may be clear as day.

You may probably cancel your gym membership if, for example, you're suddenly spending several hours each day exercising on your property.

Other things could be more subtle, requiring more discernment to determine how to lessen or eliminate them from your life.

Step 6: Learn To Preserve Food

Though there are many methods for preserving food, food preservation as a concept is kind of becoming extinct.

You may reduce your food expenses by learning even one food preservation technique, such as canning, pickling, freezing, cold storage, dehydrating, or smoking.

It's essential to understand how to preserve food if you raise your own fruits and veggies.

By the end of the season, you'll probably have much more food than you know what to do with. Furthermore, the most of it will be wasted if you are unable to maintain it.

To feed your family over the winter, you'll need to figure out how to prevent your vegetables from going bad.

Learning to preserve will enable you to purchase food in season, when it's at its most affordable and ripe, and keep enjoying it all year round, even if you don't cultivate your own food.

It's likely that you know someone who might lend you some extra canning materials to give it a try.

Even if you have to purchase a food dehydrator or canning jars, they will often pay for themselves after the first usage or two.

Finding a cool, dark spot to keep items in your basement or beneath your house is all it takes to start using cold storage.

Step 7: Make Friends With Other Homesteaders

People who homestead are often thought to either reclusive or unsociable. In actuality, however, a large number of homesteaders are amiable and willing to impart their knowledge to anybody who shows an interest.

If at any point along the process you have questions or worries, having a more experienced homesteading partner may greatly assist.

They will have firsthand knowledge of the weather, growth conditions, legislation, and a plethora of other pertinent facts, having probably gone through it all themselves.

And remember that even when everyone else is telling you that you're insane, you still need someone who shares your lifestyle and moral support.

From a material standpoint, networking with other homesteaders makes sense as well. It's simple to trade for what you need if your buddy has too many eggs and you've grown too many peppers.

Alternatively, you may establish enduring agreements with other homesteaders to exchange goods and provisions that you may not choose to cultivate on your own.

It can be more cost-effective to just borrow a plow from a neighbor rather than purchasing one for yourself if you only need it once a year at the beginning of the season.

Step 8: Start A Garden

Just plant a garden if you've read this far and you haven't already!

It doesn't have to cost a lot to garden. Actually, a few bucks will be enough to purchase a few packets of seeds. Really, all you need to get started is dirt, water, and sunlight—all of which are free.

With a little love and care, most veggies can thrive in practically any kind of soil, however, your output may not be as high as that of someone who utilizes fertilizer.

If you don't currently own any property, you may participate in a community garden or even borrow a small plot from a friend or neighbor.

In return for some free veggies later in the season, most folks are delighted to let you utilize some additional area they aren't utilizing.

Step 9: Compost

Growing food and composting go hand in hand. You will be producing wonderful, nutrient-rich soil on your own after your first year of composting, even if you cannot initially afford premium soil or fertilizer.

Throwing all of your food scraps, leaves, chicken dung, and excess plant material from your garden into the compost doesn't need much work. It's also difficult to do it wrong.

All you need to do is let it to break down and rotate it periodically to have free soil that you can use to replenish your garden.

Step 10: Learn To Sew and Mend Clothes

You'll probably start wearing out your clothes while caring for your animals or veggies while working on your farm.

You may just acquire a new pair of pants if your old ones become ripped. However, being a homesteader entails more sustainability than that!

You may fix your own garments and extend their life by months or years for the price of a little thread.

It is not required, but a sewing machine will make things quicker and easier. Clothes may be repaired and hemmed to last much longer with only a needle and thread, and you can save a lot of money doing it.

Step 11: Learn To Build and Repair

If you want to learn how to fix and prolong the life of your own clothes, sewing is a fantastic place to start. However, you should also broaden your expertise to include other trades like carpentry.

Although you don't have to become a master carpenter, you should have enough handyman skills to be able to repair items around your farm when they break without always calling someone else in.

Your fixes must keep things functioning; they don't have to be elegant.

If you can build items yourself, like tables, cabinets, or even a barn, you'll save a tonne of money.

Self-sufficiency tips

Many individuals dream of Self-sufficiency in farming, but very few actually succeed in doing it completely.

Though it takes a lot of effort, more and more individuals are attempting to take the risk.

We all have seen the effects of inflation on food and other item prices in recent years. We all have also personally seen food and gasoline shortages.

As a result, more individuals today aspire to independence.

Now more than ever, being self-sufficient is simple. But the procedure is still rather challenging and drawn out. particularly in contrast to living in an urban area and doing a traditional career.

This section will define self-sufficient farming, discuss its benefits and drawbacks, and outline the amount of land required. In addition, providing you with dozens of detailed pointers to get you going.

What is Self Sufficiency Farming?

Producing the majority of your food on your own property with little to no assistance from other sources or organizations is known as self-sufficient farming.

A self-sufficient farmer trades very little, if at all, with the outside world. They might live for years without ever leaving their farm or visiting a town since they have all they need.

It requires a great deal of preparation and discipline to be self-sufficient.

In order to survive the full winter, a self-sufficient farm, for instance, would need to stockpile adequate food throughout the warmer months.

Sustainable living and self-sufficiency go hand in hand. A self-sufficient farm makes an effort to generate as little trash as it can.

Waste materials are instead recycled or utilized in other ways. Furthermore, the farm produces practically everything that is eaten there.

To create a productive, self-sufficient farm, other ideas like permaculture and renewable energy are practically important.

To become a self-sufficient farmer, you must learn how to produce, cultivate, and market whatever you need.

How Many Acres Do You Need For a Self Sustaining Farm?

Depending on who you ask, a self-sustaining farm will need varying amounts of land.

The number of acres you need will also depend on the kinds of things you perform on your farm and your objectives.

Can you be self-sufficient on 1 acre? The response will really vary.

You can make it work if you use every square inch of an acre of land and don't mind having certain limitations on what you can plant.

How Many Acres For Vegetables?

Even 1/4 of an acre could be plenty to grow the majority of the food for a small household if you're really efficient.

In an ideal world, you could produce enough veggies in that amount of area to serve two or four people.

It's unlikely that you will be able to use all of your land. A portion of your property can be covered in forest and shaded by big trees.

Or there's a chance that certain areas of your soil are rocky, marshy, or mountainous.

How Many Acres To Raise Animals?

You may want up to 50 acres or more if you are growing livestock, particularly if you are raising bigger animals like cattle.

Every month, an adult cow requires around 4 acres of land. You can see how this will start to pile up significantly if you had a whole herd.

Don't limit your consideration to the area needed for housing your animals when estimating the amount of land required for raising them.

Additionally, after one area is wiped up, you'll need enough other pasture to rotate them to while the grass and weeds grow back.

Remember that you'll also need enough space on your property to grow enough feed to last your animals through the winter.

Do you wish to be self-sufficient and your farm is just a few acres?

You'll probably have to give up red meat and subsist mostly on hens, maybe supplemented during the year by a few pigs or goats.

Other Land Uses

Your self-sufficient farm's land will need to be used for more than simply raising animals and veggies.Energy is yet another important thing to think about.

You will need to dedicate at least a few hundred square feet if you want to use solar panels.

About 200 square feet (18 square meters) of solar panels are needed for a modest house to be completely self-sufficient.

A bigger house could need at least 1,000 square feet, or around 90 square meters.

If you live in a cold area and use firewood to heat your house, you should anticipate needing five to ten acres of forest to provide the necessary amount of wood.

Finally, consider how much room you'll need for living and storage. Most likely, your house will be at least 1,000 square feet (90 meters) in size.

In order to enter and exit the property, a driveway is also required. In addition, you should definitely have a few barns or sheds to keep all of your tools and supplies.

Sustainable living practices

How To Get Started With Self Sufficiency Farming (Step by Step)

There has never been a better moment to start if your goal is to one day own an independent farm: now. You may begin implementing some of these actions immediately, even if you presently reside in a city.

1. Get out of debt. Living off of your own resources means you won't have enough money coming in to pay off debt.

Therefore, you will need to pay off all of your credit card debt and school loan debt before you can live fully off the grid. In addition, you'll have to repay any outstanding debts for cars or other purchases.

Your self-sufficient farm should ideally not even need a mortgage since you should only be earning a few thousand dollars year.

2. Cut out addictions. Without a doubt, you should give up smoking and drinking. Unless you want to cultivate your own tobacco and make your own booze. But initially, it's best to concentrate on consuming enough food to live.

But it goes beyond drinking and drugs. Along with social media and television, you should give up other vices like your expensive morning cup of coffee from the neighborhood café.

These gradually deplete your finances or precious time that you might be using to tend to your property.

3. Get lots of exercise. You'll be doing a lot of labor-intensive, physical work on the property. You should start exercising in advance if you're noticeably overweight or out of shape.

Therefore, start immediately to become in shape. If not, you'll be worn out and run the risk of hurting yourself, which is unacceptable when your job relies on your physical well-being.

4. Start a garden. Anyone may begin here, regardless of whether they reside in an apartment. Vegetables may be started in pots in a sunny window or on your balcony.

To ensure that you have the vitamins and minerals you need on the farm, it will be essential to grow your own fruits and vegetables. Hence, you should begin studying and exploring as soon as possible.

5. Get rid of your lawn. Plant edible plants in lieu of ornamental ones. Replace shrubs with blueberry bushes and fruit trees in place of non-food producing ones.

6. Ensure you possess the necessary abilities. If you're not from an agricultural family, you have a lot to learn.

7. Obtain suitable land and water to be self-sufficient. After determining how much land you need, you must purchase the right property.

Verify if the land has a water supply. a river that flows through it or a well that may be dammed or redirected.

In the event of a drought or during the warmer months of the year, you don't want to take the chance of running out of water.

8. Buy less. Being more of a producer and less of a consumer is the key to self sufficiency. There won't be a lot of external supplies available for your homestead.

Thus, begin living a more basic lifestyle as soon as possible. Ensure that your family is on board as well, or else they could have a surprise later on.

9. Add fish to the pond on your land. Once you have a homestead, you may utilize ponds and other natural resources as an additional source of food. furthermore to increase the land's biodiversity.

10. To get all of your own meat, raise cattle. Because meat is costly, if you're living a self-sufficient lifestyle, you probably won't want to be purchasing it every week.

Find out what kind of meat is most popular in your household. Whether it's beef, lamb, hog, poultry, or another food. Next, determine the number of animals required to satisfy that demand.

11. Raise your own animals. It won't be necessary for you to purchase fresh animals each spring. Thus, you'll have to become proficient in animal breeding.

In this manner, the farm will contain animals from many generations, allowing you to build a more sustainable system.

12. Grow your own animal feed. Feeding is one of the major inputs in animal husbandry.

For them to graze, you'll need either acreage or maybe both. Animal nutrition like hay and whole grains is needed.

13. Take up hunting. There will be deer and other wild wildlife wandering about on many farms. You should understand how to use this extra resource on your property.

14. Learn to trap. Trapping is similar to hunting, but it doesn't involve spending a lot of time actively waiting about.

Catching meat for your family and getting rid of pests that are harming your crops or cattle may both be accomplished via trapping.

15. Get proficient in meat butchering. You won't be bringing your own meat to a slaughterhouse to be processed after you've grown or hunted it.

All of your meat will need processing and storage, which you must learn. You should try to avoid wasting as much as you can and save as much as you can..

16. Produce your own compost. On your farm, abandon the concept of "waste" and begin to see everything as a resource.

Unlike in the city, there won't be weekly garbage collectors picking up your rubbish.

17. Join a community. In actuality, even the most resilient among us find it very difficult to be total "lone wolves" and do everything by ourselves.

Developing a connection with your neighbors is vital. You never know when you could find yourself in a situation when you need to seek for help.

Many of them share your mentality and have accumulated years of experience and knowledge that they are willing to impart.

In order to make new friends and escape the isolation and loneliness that might come with being a self-sufficient farmer, you can also join neighborhood groups.

18. Learn to preserve food. A large portion of your fruits, veggies, and herbs will mature concurrently in the late summer and early fall.

For your family to survive the winter, you'll need to stockpile enough food. It is essential to learn how to can, pickle, and smoke food if you want to keep it from going bad.

19. Harvest your own firewood. If you want to use a wood burner to heat your house, you must own and maintain at least a few acres of woods.

You should budget at least a couple full days a year for the preparation and piling of firewood.

20. Install renewable energy. You're gonna have to live off the grid in order to be completely self-sufficient.

That does not imply that you have to give up the ease and comfort that come with power. However, in order to collect and store it yourself, you'll need to put up solar panels, windmills, or other sources.

21. Take out of the grid all of your water. You should, at the very least, drill one or more wells on your land to collect groundwater. Ideally, there should be a river, stream, pond, or other water source nearby.

22. Collect rainfall. Adding to the previous suggestion, you may increase your water supply by using rainwater collection. It may be used for various things, such watering crops.

However, depending on your location, you most likely won't be able to subsist all year long on rainwater alone and would want another source.

23. Install a composting toilet. Every three to five years, your property's septic system has to be emptied and maintained.

Investigate building composting toilets or other more self-sufficient and sustainable options instead.

24. Put permaculture to use. It is preferable to cooperate with nature rather than fight it in order to be really sustainable.

25. Grow mushrooms. Protein, amino acids, and other nutrients that are difficult to get in plants may be found in abundance in mushrooms.

They are a great addition to your diet, or they may even take the place of meat.

It's unlikely that many other farmers or homesteaders in your region are cultivating them. They are thus a useful item to trade with.

26. Acquire woodworking abilities. Many wooden items on the farm, such as fences and barn doors, may need to be fixed as they break. In addition, you make the majority of the furnishings for your house.

27. Milk cows and goats. You must have animals that generate milk and cheese if you like to consume either of these foods.

How many cows or goats you need depends on whether you want to create a microdairy to generate extra money or whether you simply want some milk as a special treat every now and then.

(28). Keep bees. You may consume the honey that bees create. However, they also yield beeswax, which is used in the production of candles, soaps, and other goods.

29. Conserve seeds and cultivate heritage cultivars. To raise your own food, you don't want to depend on seed corporations every year.

At the conclusion of the season, you will need to conserve some of each plant's seeds for the next year.

When initially starting out, be careful to use organic heritage kinds of seeds.

Any hybrid seed, even F1 plant seed, cannot be stored for use in subsequent years and will not produce true-to-seed.

30. Consider cutting your hair by yourself. Living in a metropolis makes you take many little things for granted, such as haircuts. You should learn to cut your own hair and the hair of every member of your family when the closest barber or hairdresser is an hour or more distant.

31. In what ways can you become self-sufficient? A staple crop of some kind should be at the top of your list.

- Typically, this crop is high in calories and carbs. similar to potatoes, rice, maize, or wheat.
- All of these crops also have the ability to be kept for an extended period of time.
- Vegetables, such as salad greens, are rich in vitamins and minerals. However, they don't provide the precise number of calories required for field labor.

Thus, the majority of your meals must include these essential foods that are high in energy.

32. Learn to fix things yourself. Most farmers become jack-of-all-trades or handymen out of need.

It's not always possible to call in an electrician or plumber. Furthermore, it would be expensive to do this for each little issue you had. As a result, you'll need to experiment a lot.

33. Make your own soap. Both the components and the procedure of making soap are quite simple. You'll need a means of cleansing. Plus, you'll be buying one fewer item if you make your own soaps.

34. Grow your own herbs. Eating potatoes and plain wheat for every meal could become monotonous. To add some taste, you'll need a few spices. You could also wish to cultivate a few hot pepper plants if you're a spicy person.

35. Cook from scratch. When you are a self-sufficient farmer, there is no food delivery service. You will thus prepare almost all of your meals at home on the ranch.

36. Sew and mend your own garments. It's likely that the farm won't have a loom for you to make your own thread. Thus, you will occasionally need to purchase cloth or barter it.

But you can make your clothes last longer by learning to sew, hem, and fix them yourself. A little hole in a pair of jeans doesn't warrant discarding them.

Conclusion

Homesteading is becoming more and more popular every year, and many people consider it to be a romantic and perfect way of life. It's not for everyone, however.

However, if you're prepared to work hard and put in the effort, it may be a really fulfilling way of life.

You don't have to live on a farm to begin homesteading; it's a way of thinking and doing life.

You may approach your ultimate objective of being a homesteader step-by-step by progressively becoming more self-sufficient and simplifying your life. Once everything is set up, you can start homesteading and earn money.

If you choose the correct balance of short- and long-term projects to guarantee a consistent flow of income for your homestead, homesteading may be financially rewarding.

Operating a homestead is similar to operating any other kind of company in terms of earning money. In the end, the choices you make and the effort you put in will determine whether it succeeds or fails.

Not every homesteading endeavor will turn a profit in the first year. For instance, it will take at least five years for your seedlings to begin bearing fruit if you are growing an apple orchard.

The homestead with the best potential of long-term success is the one that is diversifying.

Earning $1,000 to $2,000 a year for a hobby doesn't seem like much. However, it soon mounts up when you have many side businesses on your farm that each bring in several thousand dollars.

Even a tiny homestead may be quite lucrative with the right methods and setup.

Why Do People Homestead?

Homesteaders may be a varied group who may not all have the same ideals or motivations for homesteading. Some are probably retiring from well-paying careers that have given them the means to purchase the necessary infrastructure for properly sustaining themselves on the land. Some people could be starting from scratch when they homestead, building a tenacious fortress to support themselves in the event of financial difficulty. Despite the fact that these two scenarios could hardly be more unlike, both individuals identify as homesteaders.

In one way or another, these folks desire to return to the land. Maybe they are fed up with the disconnection between physical work, food production, urban life, and the seasons. Perhaps they want to give up the "rat race" and live a calmer, more straightforward lifestyle. They may choose to pursue careers in farming or gardening, animal care, or manual labor. There are a wide range of motivations for wanting to homestead, and it is undoubtedly a very fulfilling effort.

BONUS

How to Be Self-Reliant: Mindset, Skills, & Gear

BOOK 7

WILD EDIBLES: A FORAGER'S HANDBOOK

Foraging may end up being your new favorite pastime if you're interested in leading a more environmentally friendly lifestyle. By doing this, you may increase your nutritional intake, explore new and intriguing foods, and even lessen your carbon impact.

Having stated that, before you begin, you must familiarize yourself with foraging safety.

This book gives an overview of foraging, including a list of frequently foraged items, and offers advice on how to begin foraging in both urban and rural areas.

What is foraging?

Historically, people have obtained their food by either hunting and fishing or by collecting edible plants, berries, and seeds from the wild.

These days, we get our food in a totally unusual manner. The majority of individuals in developed nations like the US get their meals from supermarkets or food delivery services.

You may not be very involved with or connected to the food you consume, other than putting it in your shopping basket, cooking it, and eating it.

But there's a rising environmental movement that promotes supporting local farmers and producing your own food. The practice of foraging, or looking for food in the outdoors, has become quite popular, particularly among those who want to live more sustainably.

Contrary to popular belief, foraging is also feasible in urban settings like cities. It is not limited to natural places like woods.

This is due to the fact that even the most metropolitan areas include parks and yards—green spaces—where edible wild plants may thrive. Edible plants such as berries, greens, and mushrooms are abundant in rural and wilderness locations.

Foraging might be appealing to you for a number of reasons, such as enjoyment, a desire to be more in tune with nature, or the health advantages of consuming locally foraged foods.

Foraging is looking for edible wild plants in both rural and urban settings, such as greens or mushrooms.

What foods can you find while foraging?

Numerous wild foods are often sought for by foragers. While some focus on locating certain items, like mushrooms, others gather all edible wild plants in their vicinity.

These are a few of the most popular meals that foragers tend to seek.

Mushrooms

With foragers, mushrooms are quite popular.

Many wild mushrooms, such as oyster mushrooms (Pleurotus ostreatus) and hen-of-the-woods, also known as maitake (Grifola frondosa), are very nutritious and safe to eat.

A thorough understanding of mushroom identification is necessary for mushroom foraging since many toxic wild species might be confused for edible ones. Therefore, it's essential to mushroom forage with a knowledgeable forager who can safely identify edible varieties.

Greens

In the wild, edible greens are widely available and may even grow in your garden. Indeed, wild greens that you may add to delectable recipes are what some people believe to be weeds.

Some edible wild greens include wild lettuce, mallow, dandelion greens, sweet fennel, plantain, purslane, lamb's quarters, and chickweed.

Notably, wild greens are rich in many nutrients.

A research that examined wild greens foraged in California discovered that one cup of mallow (Malva sylvestris) had 27% more calcium than the same quantity of whole milk, while one cup of dock (Rumex crispus) surpassed the recommended adult diet of vitamin A.

The research also revealed that the examined wild greens were typically more nutrient-dense than kale, with the exception of vitamin C.

Berries and fruits

In many parts of the United States, berries and other wild fruits like grapes and pawpaws are available.

Berries that grow wild include blackberries, raspberries, blueberries, cloudberries, cranberries, bilberries, currants, lingonberries, bearberries, and crowberries.

Research indicates that wild berries and other edible wild fruits are very nutrient-dense, offering a range of antioxidants and anti-inflammatory substances that might be good for your health.

Other commonly foraged foods

While berries, greens, mushrooms, and other fruits are among the most often consumed foraged foods, there are many more edible wild plants that may be harvested and consumed. Depending on where you live, you may forage in the wild for roots, nuts, seeds, and even shellfish.

Certain coastal locations provide opportunities for collecting shellfish such as mussels and clams, which are great sources of several nutrients.

In addition, a lot of foragers gather wild onions and the roots of chicory, dandelion, and burdock. You may add these nutrient-dense roots to a range of recipes.

Acorns, walnuts, pecans, and pine nuts are just a few of the nuts and seeds that some foragers like gathering.

Identifying edible wild plants

Plant species that are safe for ingestion by humans are known as edible wild plants. Understanding which wild plants are edible may be a lifesaver, a wonderful way to engage with your community's ecology, and a way to add some diversity to your daily diet. Edible wild plants may be found in almost every habitat. Many weedy edible plants thrive in close proximity to populated areas. Others are found growing in mountain meadows, deep in the woods, among streams, or in the forest's understory. Investigating the kinds of edible plants that are present in your region is the best method to start foraging. As always, the finest plants to forage are those that haven't been impacted by adjacent highways or human activities.

Nutritious Wild Plants

In the wild, there are hundreds of kinds of edible plants. contemplate about the following typical categories of consumable plants:

1. Acorns (Quercus spp.): The nuts of the oak tree are called acorns, or oak nuts. In North America, there are over a hundred different kinds of oak trees, and many of them provide food for both people and animals via their seeds. Acorns are hard shelled, with a part on top that looks like a little cap. The acorns may be made safe to eat by leaching, which is the process of repeatedly soaking them in hot or cold water to eliminate the bitter tannins. Some species of oaks that are edible to humans include White Oak, Black Oak, Pin Oak, and Red Oak.

2. Burdock (Arctium spp): If you come into contact with the thistle-like flower pods on this plant, they will adhere to your clothes. Burdock has edible roots, leaves, flower stalks, and blooms. The root may be roasted and eaten like other root vegetables, or it can be used to produce an anti-inflammatory tincture. The leaves may be cooked into a variety of meals or eaten raw.

3. Cattail (Typha latifolia): Native American cultures have long used cattails—reeds that grow in streams throughout North America—for a variety of uses. Consuming cattails gives you a good amount of potassium, phosphate, and vitamin C. At various periods of the year, different portions of the plant are best consumed. For instance, spring is the finest time to enjoy the new shoots, flowers, and pollen, while autumn is the best time to enjoy the stalks and roots.

4. Chickweed (Stellaria media): This wild green's leaves, stalks, tiny white blooms, and seeds are all edible. During the colder months, chickweed develops in dense, luxuriant stands; by late spring, the plants dry down and go to seed.

5. Curly dock (Rumex spp.): Thick, deep-growing roots of this wild edible green are used both medicinally and for food. The greens may be eaten raw, or the seeds processed into a flour similar to buckwheat.

6. Dandelion (Taraxacum officionale): One of the most widely available edible wild plants, dandelion may be found in backyards and woodland meadows alike. The dandelion's blossom, roots, stems, and leaves are all edible. Dandelion greens are somewhat bitter

and peppery, like arugula, but they are also significantly longer. When the blossoms and leaves develop in late winter or early spring, you may harvest them and use the leaves and blooms in salads and sautéed foods. The roots can be used to brew tea.

7. Elderberry (Sambucus canadensis): The tiny, dark purple or black berries of the elderberry shrub are called elderberries. Cooked elderberries are a common addition to wine, cordials, tea, pies, and other baked items. They are also used to make jams and jellies. While cooked, ripe elderberry fruit and blossoms are edible, most other components of the bush, such as raw or unripe berries, leaves, stems, and roots, are mostly poisonous and may make you sick if you eat them.

8. Garlic mustard (Alliaria petiolata): Garlic mustard has slightly bitter, mustard-like leaves. Garlic mustard has an aroma is similar to garlic, hence the name. It is safe to eat this plant when it is young—you can consume older plants after thoroughly cooking.

9. Jerusalem artichoke (Helianthus tuberosus): The tuberous roots of Jerusalem artichokes, also referred to as sunchokes, may be eaten raw or cooked. These artichokes feature large, ovoid, somewhat hairy leaves and sunflower-like, rich yellow blooms.

10. Lamb's quarters (Chenopodium album): Lamb's quarters, often referred to by the popular name "pigweed," resemble wild quinoa, which yields edible grains if allowed to go to seed. The leaves may be used to create a tasty and nourishing wild salad. They may also be processed to form a wild food pesto, or sautéed like spinach or Swiss chard.

11. Miner's lettuce (Claytonia perfoliata): An edible annual herbaceous plant is called miner's lettuce. Winter purslane, or miner's lettuce, is a plant that grows in shaded regions of North America's coastal and western highlands. Native Americans have been eating it for millennia. The popular name "miner's lettuce" comes from the California Gold Rush miners who were looking for a way to prevent scurvy by consuming vitamin C. It is still a very common forage plant since it is pleasant, nutritious, and simple to obtain.

12. Morel mushroom (Morchella spp.): These are oblong mushrooms with a pocked, almost honeycomb-like top. They are a delicious wild mushroom. Foraged in recently burnt regions, morel mushrooms are among the first species to restore a habitat after a wildfire. Like any mushroom, there are dangerous lookalikes, so it's best to check with an expert guide.

13. Pigweed amaranth (Amaranthus spp.): These wild plants have edible leaves, flowers, seeds, and flower buds. The tenderest and most delicious leaves are young ones.

14. Plantain (Plantago spp): Edible seed pods and leaves are characteristics of the broadleaf plantain. When the taste and nutritional profile of the leaves are at their finest, they are best consumed young. Plantain leaves make a nice spinach alternative, albeit being a little rough.

15. Stinging nettle (Urtica dioica): This wild green that is edible grows in thick stands of serrated leaves that are coated in small, irritating hairs. Gather them carefully, and then pick them up with strong gloves before putting them in a bag or container. Once cooked, nettles become excellent, healthy greens that work well in casseroles, soups, and pasta dishes. They also no longer sting. The leaves may also be steeped to form a nutritious tea.

16. Watercress (Nasturtium officinale): Shallow, flowing water is ideal for watercress growth. The flowers, stems, and leaves are edible and have a flavor that is a little bit spicy. Make sure the water and the surrounding region are clean, just as you would with any wild edible.

17. Wild garlic (Allium vineale): This plant has edible flowers, leaves, bulbs, and bulbils. The taste and scent of wild garlic are comparable to those of cultivated garlic. The bulbs and bulbils may be eaten on their own, pickled, or as a snack with other dishes.

18. Wild onion (Allium bispectrum): Six-petaled, white and purple blooms adorn wild alliums. The plant's edible components include the bulb that resembles an onion and is where it grows. They smell like onions, which sets them away from other potentially deadly lookalikes like Lily of the Valley. The flavor is milder than onions found in the grocery store.

19. Wood sorrel (Oxalis spp.): Wood sorrel is a common edible plant that also produces delicious yellow flowers. Its leaves are similar to those of clover. Their unique citrus taste makes them a wonderful complement to a wild greens salad.

To find out whether any berries and plants you encounter in the wild are probably safe to eat, use the edibility test. The procedure of the test involves a number of phases that progressively expose your body to more plant contact.

Universal Edibility Test

How to Identify Edible Wild Plants

Food that has been freshly foraged may be enjoyable to harvest and, in times of scarcity, may even save your life. But the most crucial aspect of foraging is safety, which is why an edibility test is necessary.

When foraging, the universal edibility test is the most reliable method of identifying edible wild plants. You will chew, cook, and swallow little amounts of a plant during the duration of the test, and you'll be watching to see how your body responds. Take these actions:

1. Seek for the most prevalent toxic characteristics. When foraging in the wild, steer clear of any plant that has characteristics with hazardous plants. Discard mushrooms and any plants that have umbrella-shaped blooms, fine hairs, spines, or glossy, waxy leaves. Generally speaking, green and white berries indicate an inedible plant. Seek for a significant amount of a single plant, since this will indicate that it is a safe food source once you have completed the extended edibility test.

2. Separate the plant. It is not true that every component of the plant is edible just because some are. Sort the plant's stems, roots, leaves, and flowers, then decide which section to test. When taking a risk in the wild, always do extensive study and test portions of the plant individually before swallowing the whole plant.

3. Test for skin contact. Apply the portion of the plant you want to consume on your outer lip, inside forearm, or inside elbow. Give it 15 minutes. You may proceed with the remainder if there are no tingling, burning, or other side effects.

4. Do a taste test. After five more minutes, taste the same area of the plant. If you experience any burning, tingling, or other uncomfortable feelings, spit the plant out and rinse your mouth with water. You may proceed with the exam if not.

5. Conduct an expanded taste test. If you do not experience any numbness, soapy flavor, or bitterness, take a teaspoon of the same plant material and chew it for five minutes, often spitting out additional saliva. After swallowing, give it eight hours.

6. Eat a small amount. Eight more hours should pass before consuming one tablespoon of the identical plant portion if you're still not having any stomach problems. You may consider that portion of the plant edible in the way it was cooked if you are still not experiencing any symptoms.

7. Exercise cautious. Use common sense and your best judgment while looking for edible wild plants, whether you're engaging in sustainable foraging or you're just attempting to eke out an existence in the outdoors. Certain plants that are edible may have dangerous counterparts or even edible berries with poisonous stems and bark (like elderberries).

Edibility Test Tips

Evaluating a plant's edibleness is simply one of many abilities that will come in handy in a survival scenario. Here are some pointers for doing a universal edibility test, regardless of whether you're in a wilderness survival situation or are just enjoying some fun foraging.

1. Perform it while you are starving. Eight hours before to doing the edibility test, eat no food and plenty of water. Avoid eating anything other than the portion of the plant you are evaluating while doing the test.

2. Compile an abundance of plant components. It takes at least sixteen hours to finish an eligibility exam. After going through this laborious procedure, you don't want to find that you're missing out on enough of the plant to satisfy your appetite. If the plant is edible, collect plenty of it before doing the test to ensure you have enough to eat.

3. Ignore it if you're unsure. Certain traits seen in deadly plants, such as prickly hairs or discolored sap, may also be present in edible plants. Even while not all plants with these characteristics are dangerous, it is still not worth the risk to suffer from these plants' negative consequences.

4. Steer clear of plants that smell like almonds. An almond smell is a natural marker of toxicity in plants (typically cyanide). Berries or plants with an aroma similar to burned or raw almonds should not be consumed.

5. Examine the leaves. Plants that resemble parsley, dill, or parsnip greens should be avoided since they are probably harmful.

6. Believe in your taste buds. There's a reason why toxic plants taste unpleasant; they're not meant to deceive you into poisoning yourself; they're simply attempting to defend themselves from pests and bugs. Spit out the plant part you're trying right away if it tastes harsh or soapy.

Foraging ethics and safety

Proper Identification of Wild Edibles

Make sure a wild plant is safe to consume before consuming it.

Locate a mentor. Gaining knowledge from a specialist or someone with greater expertise can boost your self-assurance.

Grab a Good Book. Although a decent field guide can't replace a mentor, it's a close second. As your comfort level with foraging increases, a reference book will boost your confidence. Not only may books be used to positively identify plants, but they are also an excellent resource for discovering new plants in your region that you may not have yet discovered. It gives you motivation to explore new areas in search of wild foods.

Before going out into the woods to forage, familiarize yourself with the few harmful species in your region. You may forage for the edible species with more ease if you are aware of the potentially toxic plants that you may come upon.

Don't depend on common names all the time. Common names may be used to identify a variety of plants. There are edible wild plants with familiar names that are also toxic. Latin names have more credibility. When offered hemlock tea, for instance, you can inquire as to whether it is infused with Conium maculatum, often known as poison hemlock, or with the delicate tips of Tsuga canadensis, also known as the eastern hemlock tree. Since Latin is a dead language and cannot change, it was selected to categorize plants and animals, while common names are dynamic and may change over time.

Utilize all of your senses. Take use of more than just visual ID. The variations of many edible wild plants are similar. Acquire the ability to differentiate between similar plants by their touch, feel, smell, etc. Toxic plants normally taste or smell bad, however this isn't always the case. Taste should only be used, however, if you are certain the plant is safe. Some plants, such as water hemlock, may be lethal in very little doses..

Study the habitat. You will not find ramps in a marsh, nor cattails on a steep hill.

Study up on companion plants. Many plants may often be seen growing close to other species. There is a significant possibility that pokeweed will be nearby if you encounter yellow dock.

Acquire the skill of identifying edible wild plants in any season. There are two reasons why this matters. Positive identification comes first. Finding perennial plants to harvest in the early spring is another incentive to monitor wild edible plants through the seasons. For instance, pokeweed is often beyond its prime by the time it can be identified. When it first emerges in spring, you'll know where to locate it if you take notice of it over the warmer months.

Find out the portions of a wild plant that may be eaten safely. Not every portion of a wild plant is edible, even if it is deemed edible. For example, the bark, stems, and roots of elderberries are deadly, but mature, cooked berries are safe to consume. It's also crucial to remember that certain plants are edible only during specific seasons. For instance, after stinging nettle grows to seed, it should not be utilized.

Maintain a foraging notebook. This is crucial for learning what is and isn't available in your region at any given time. By meticulously documenting your foraging discoveries over many months or years, you will progressively create a calendar that indicates when harvesting opportunities will arise. This will assist you in creating a meal schedule in advance as well.

Sustainability

Avoid overharvesting. All populations have their limits. The colony should be respected even in situations where there are a lot of wild edible plants. Aim to gather no more than ten percent (or less, depending on the level of pressure from foraging in the region). Naturally, you should never gather more than you intend to utilize.

Steer clear of foraging for rare and protected edible wild plants. Though they may be common in some areas of their habitat, many plants are uncommon elsewhere.

Gather just the portions of the plant that you want to use. Remove the sapling before processing the leaves of the sassafras to create file powder. Just take what you need, leaving enough behind to guarantee the plant's continued health. If you do not need the whole plant, a decent rule of thumb is to pick no more than 25% of the plant.

In your garden, think about growing wild edible plants. Edible wild plants are easily transplanted and multiplied. For example, ramps are becoming less common as a result of overharvesting, but they may still be grown under the correct circumstances. Spend some time learning about local growth environments for uncommon plants. Interest in gathering edible and therapeutic plants from the wild is growing again. Wild plant populations are under extreme strain due to this and habitat degradation. It puts the onus of preservation on us.

Safety

Avert risky places. Avoid foraging for edible wild plants close to busy roadways. Lead and other heavy metals are absorbed by most plants from hazardous exhaust. Furthermore, even when the traffic stops, these poisons often find their way into the ground. Additionally, stay away from locations where pesticides are or have been used.

Determine which plant parts are safe to touch at each season. Although it was already said, this is nevertheless important to note.

Be aware of the water source while gathering wild water plants. This is particularly crucial if you want to consume the wild edible uncooked. It is the same as drinking polluted water when you eat plants that have grown in contaminated water. Cooking does not eliminate the risk of chemical and heavy metal contamination.

Use only plants that seem healthy for foraging. Plants may be affected by pollution, pests, fungus, or diseases. Harvesting healthily reduces your chance of getting sick and also results in more nutrient-dense produce.

Ask before you go foraging. Although there may not be a clear safety concern here, breaking the law and property rights might have some very severe repercussions. It's also just politeness.

Preparing wild edibles

You may be wondering what happens when you locate edible delicacies like mushrooms and ramps, and I'm sure you've heard others getting thrilled about them. You may make the most of your wild edibles by following the instructions in this section. You'll become an expert at using wild foods in no time with these culinary methods!

CLEAN AND WASH THOROUGHLY

Once you've finished your foraging, thoroughly clean your wild plants and food to get rid of any dirt, bugs, or trash. After harvesting, be sure to take care of this as quickly as possible. The difference between eating properly and wilted, slimy vegetation is as simple as leaving it out for a day.

START SIMPLE

Start with easy dishes that bring out the flavors of the foraged food. By doing this, you may enjoy every plant's distinct flavor to the fullest without adding too many elements to the meal.

TAKE A TINY TASTE TEST

Taste a little piece of a new wild plant before incorporating it into a dish to make sure it lives up to your expectations for flavor, texture, and possible bitterness. To get rid of any bitterness, certain wild plants may need to be blanched or cooked.

USE IN FAMILIAR DISHES

To ease the adjustment, include foraged produce into meals that you are already acquainted with. Use wild berries in baked products, wild mushrooms in pasta dishes, and wild greens in salads. Using wild edibles in cooking may help you recognize their tastes in the context of foods you like.

PRESERVE THE HARVEST

If you find yourself with an abundance of foraged food, think about canning it for later consumption. Berries may be frozen, jams and preserves made, herbs dried, and vegetables pickled. You may continue to enjoy foraged food beyond the foraging season by using conservation practices, which prolong its shelf life.

EXPERIMENT WITH COOKING METHODS

Try a variety of cooking methods to get the full taste of foraged foods. Try roasting, grilling, sautéing, or even adding them to stews or soups. You may learn the optimal techniques for enhancing and preparing the tastes of diverse wild plants by experimenting with different approaches.

When adding foraged food to dishes, think about combining it with tastes that go well together. For instance, balance bitter vegetables with creamy sauces or cheeses, or mix sour berries with sweet foods. To make well-balanced recipes, try experimenting with seasonal items and taste combinations.

CONNECT WITH COMMUNITY

Exchange experiences, recipes, and expertise with other foragers and food aficionados. Participate in online forums or neighborhood foraging clubs to share advice, recipes, and firsthand knowledge.

DOCUMENT YOUR DISCOVERIES

Record your discoveries of wild plants, as well as any recipes or culinary methods you've tested, in a diary or digital database. This makes it easier to keep track of your progress, remember tried-and-true recipes, and record unusual discoveries made while foraging.

Embrace the process

Be willing to change and grow. Using wild plants for cooking and foraging is an ongoing educational process. Accept the journey, grow from your errors, and acknowledge your accomplishments. With each encounter, you get greater self-assurance and expertise while handling foraged food.

Savor exploring new tastes, preparing with wild foods, and integrating the abundance of nature into your culinary endeavors!

DOCK, DANDELION AND NETTLE SPRING PUDDINGS RECIPE

This dish involves a good deal of fuss, much of which may be skipped if you want a more rustic result (such as blanching and pané). The mixture might also be presented beautifully as the centerpiece of a buffet in place of the laborious task of rolling it into small balls. However, there is a certain pleasure I get from investing the time and energy necessary to transform these ordinary weeds into a sophisticated, savory "truffle" fit for any upscale restaurant menu. Amazing things don't always have to be uncommon!

Ingredients

Makes about 35 mini puddings or one clonker

- 400g pearl barley
- 8 eggs
- Half a carrier bag of wild leeks or 2 large cultivated leeks
- About half a carrier bag of cleaned dock and dandelion leaves, or any other spring green (plantain, hogweed shoots, watercress, etc.) with a fairly strong flavor. Wild garlic is another option, however cooking them does cause them to lose most of their pungent flavor
- Ice

- A handful of whatever fragrant wild herbs you choose, finely chopped (optional); wood avens' roots are still flavorful and cow parsley is rather prevalent at this time of year.
- Optional finely ground dried seaweed. As seasonings or umami, laver (aka nori), gutweed, pulse, sea lettuce, and pepper dulse are all excellent choices.

For the pané (optional)

- Four tsp finely chopped dry bread crumbs - if, like me, you struggle to save breadcrumbs, try pulverizing some breadsticks.
- 1 egg
- Four tablespoons of dried and ground nettle (optional) may be used to make pané; a dehydrator makes this simple, or you can bake it at a very low temperature. In the event that it doesn't work, most health food stores carry it.
- Flour
- Vegetable oil for frying

Method

1. Once the pearl barley is soft, boil it, then drain and set aside to chill.
2. Hard boil four eggs, then finely chop.
3. Finely chop the leeks and cook them in little oil until they become tender.
4. Refresh with cooled (or very cold) water after blanching your half carrier bag of greens in boiling water for 20 seconds. You may skip this step, but the resulting pudding(s) would taste less colorful and more bitter.
5. Finely chop the blanched greens after squeezing out all the water. Your mixture will not bond together as well if you leave them damp and mushy.
6. Combine three raw eggs with the chopped eggs, leeks, barley, chopped greens, aromatic herbs, and powdered seaweed, if using.
7. Liberally season with salt and pepper.
8. Transfer mixture to a big tray, completely cover with cling film, and set the tray in a larger tray of boiling water (bain-marie). Alternatively, use the method outlined here to encapsulate the mixture in a bowl using foil, greaseproof paper, and thread. Make sure the bowl isn't in direct touch with the pan's base by submerging it halfway full of water in a pan (an upside-down saucer works nicely for this).

9. Steam the mixture on the stove or in the oven for 45 to 60 minutes at 160° C. When it comes out, it ought to hold together well enough to roll into balls. If not, combine and continue to steam for an additional one or two raw eggs.

10. Spoon onto a platter and serve if serving as a big pudding.

11. Roll the mixture into little balls (this may be difficult and takes a delicate touch; it's better to roll while warm) and refrigerate if serving as mini-puddings. At this point, the mini-puddings may be frozen, but be sure to freeze each one separately before bagging. At this point, you may simply flour and fry them if you don't want to deal with the pané.

12. Combine the finely ground dry bread crumbs with the dried, blitzed nettles.

13. Use a fork to gently beat the raw egg.

14. After dipping the cold balls in flour, beaten egg, and bread/nettle mixture, deep fried them in vegetable oil for one to two minutes, or until the pané is a pleasing shade of brown.

15. Present warm. They pair very well with fermented wild leeks, deep-fried reindeer moss sprinkled with powdered cep, or any other finger-food that naturally pairs well with wild ingredients.

Conclusion

You'll become better at foraging the more you do it. Every time you go foraging, try learning about a new wild edible plant. Learn about all of a plant's applications, including therapeutic ones. As you expand your list of practical wild plants, you'll become more at ease with the natural world and all it offers.

Getting Ready for Trips into the Wilderness

There is a higher chance of severe harm while engaging in several outdoor activities. Extensive survival equipment, such as food, water, maps, protective clothes, first aid supplies, and mental and physical toughness, are necessary in wilderness situations. This book is meant to serve as a reference and instructional tool; it is not meant to replace practical knowledge and experience.

Happy foraging!

BONUS

Easy Edible Plants for Beginner Foragers- Eating Wild Food

BOOK 8

CULTIVATING ABUNDANCE: A GUIDE TO GROWING YOUR OWN VEGETABLE GARDEN

You may question, why garden? Would you want to taste the greatest fruit and veggies you've ever eaten? The sweet, juicy tastes and colorful textures of food straight from the garden will wow you if you've never experienced them. Nothing compares to fresh vegetables at all, particularly if you cultivate them yourself, which is possible!

Though it might be intimidating at first, gardening is a very fulfilling pastime. This guide will go over the fundamentals of vegetable gardening and planning, including how to choose which veggies to produce, how to design the ideal garden size, and where to put your garden.

Vegetable

The word "vegetable" is used to describe a wide range of plant components, including their leaves, roots, fruits, and seeds. Globally, vegetables constitute a staple diet and an essential component of contemporary agriculture.

Vegetables are abundant in nutrients and low in calories, so most health experts advise consuming them every day. The scientific community has come to the conclusion that one of the greatest methods to start getting nutrients from food at a young age is to eat a balanced diet that rotates among various kinds of vegetables.

Health Benefits

Vegetables are a great source of critical vitamins, minerals, and antioxidants that are good for your body in many ways. Carrots, for example, are well recognized for having a high vitamin A content, which is crucial for maintaining eye health as you age.

There are several more health advantages of vegetables, including as:

Improved Digestive Health

Dietary fiber, a kind of carbohydrate that facilitates the passage of food through the digestive system, is abundant in vegetables. According to studies, fiber may also enhance the body's absorption of vitamins and minerals, which may increase your level of energy throughout the day.

Lower Blood Pressure

Potassium is found in a lot of green leafy vegetables, such as kale, spinach, and chard. Potassium may lower blood pressure by assisting your kidneys in their more effective removal of salt from your body.

Lower Risk of Heart Disease

Vitamin K, which is also included in green leafy vegetables, is thought to stop calcium from accumulating in your arteries. By doing this, you may lessen your chance of artery damage and avert a number of potential heart health issues.

Diabetes Control

Vegetables have a high fiber content, which is essential for good digestion. Because of their low glycemic index, they won't cause your blood sugar to spike right after a meal. The American Diabetes Association suggests eating non-starchy veggies including cauliflower, broccoli, and carrots in three to five servings a day.

Nutrition

Folate, a B vitamin that aids in the production of new red blood cells in the body, is abundant in vegetables. In addition to being particularly beneficial for children's health, folate may also lower the chances of depression and cancer.

Vegetable gardening essentials

You are presumably planning to get started with a new vegetable garden. Congratulations! I assure you that this is the finest choice you will ever make.

Before you create the groundwork for your new garden, there are five very critical things you should know. If you do things correctly, you should have plenty of harvests.

1. Get the Location Right

You must choose the ideal location for your vegetables if you want to get the most out of producing your own tasty and nourishing foods. In most cases, compromises are unavoidable, but the same guidelines hold true regardless of the size of your new profitable plot.

Light is the first thing to take into account. Most crops need six to eight hours of direct sunlight, especially in colder, temperate settings like mine. Throughout the day, note any areas that are shaded. Naturally, you are unable to move buildings or other items outside of your control, but you may take action against overhanging branches or trees within your boundaries.

In contrast, if you garden in a hot area, you may deliberately want to look for a spot that gets some shade so that you can plant cool-season vegetables like spinach and cabbage in the warmer months.

Remember to account for buffeting and the predominant winds. The best spot will be one that is protected from the worst winds and has at least some sunshine throughout the day.

Soil is next. While there are few gardens with ideal soil for growing vegetables, you should steer clear of places that are very wet or poorly drained. You may also think about growing on raised beds, which will assist to keep the roots of your plants above the soggy soil.

2. Plan Your Garden's Layout

After deciding on a basic site, the garden's actual layout must be taken into account.

The finest beds are often narrow ones. Whether they are in the ground or on raised beds, they should not be wider than 4 feet (1.2 meters). Since you shouldn't ever have to walk on the soil, this helps to prevent compaction and facilitates crop rotation and upkeep. Create spaces that are both readily tended to face-on and let a wheelbarrow to pass through, carrying loads of mulching compost, for example. At least 18 inches (45 cm) of route width is recommended.

In the heat, watering may be quite the task. Locating your vegetable garden next to a water supply makes irrigation much simpler. If that's not feasible, you may install rain gutters and water collecting barrels on surrounding sheds, greenhouses, or other buildings. To ensure you have a enough supply of the finest water for plants, install as many water barrels as you can. Purchasing a premium hose with a good spray attachment is also worthwhile.

You should also include a space for composting (more on that in a moment) and a place to relax and take in the beauty of your newly created garden since it's crucial to sometimes take a moment to appreciate your labor of love!

3. Clear the Weeds

Once you've chosen your site and layout, the next thing you need do is get started on your beds and growing areas so that you'll be prepared to plant when the time comes. This is a terrific way to gain ahead in the winter, when growth has slowed to a crawl.

Before beginning to tackle the weeds, the bigger boulders and debris will need to be cleared out. You must control weeds because they may compete with your crops for nutrients and moisture.

Covering weeds is the simplest approach to eradicate them. It's that simple: weeds will ultimately run out of resources and die if light cannot reach them. For the most part, all that has to be done is trim the grass or weeds close to the ground, cover with cardboard, and then add at least 4 inches (10 cm) of organic materials on top.

The annoying, unpleasant perennial weeds are those! despite the fact that they take a while to die down, you may still utilize cardboard and organic stuff. Just hoe or pluck away any weed shoots that breakthrough, and the roots will finally succumb. A more comprehensive approach would be to completely enclose the area and keep it covered with a light-blocking substance, such black polythene sheets, for as long as feasible to significantly weaken and, ideally, eradicate the weeds.

Warmer weather causes weeds to grow more quickly, which wears them out sooner in the growing season. Nevertheless, because there is more time during the winter, it makes sense to install coverings.

4. Start Improving the Soil

The sooner you work on improving your soil, the longer it will take for it to become suitable for planting and seeding. Though very few soils are ideal for producing vegetables, almost all soils may be made better by adding plenty of lovely, rich organic matter, like leafmold, well-rotted manure, or garden compost.

Adopt no-till gardening techniques. There are several benefits to this. The purpose of leaving the soil undisturbed is to allow the complex web of soil life to flourish and encourage better crop development. Weed seeds remain buried when the earth isn't dug up, which results in less weeds. Naturally, not excavating also saves a ton of time and work! It works well with the smothering approach of controlling weeds since, most of the time, you can plant directly into the compost on top of the cardboard without having to wait for the weeds to die.

Spreading organic matter on top of already-weed-free soil and allowing the worms to "dig" it in for you is a simple way to improve the soil when establishing new no-till growing areas. Now is the time to get your organic debris onto growth regions so the worms can begin to work.

It's astonishing how much organic matter a successful vegetable garden needs, therefore it pays to produce as much of your own as possible. Getting your composting area set up is really important! Having two compost heaps is ideal as it allows you to actively add organic matter to one while letting the other grow.

Your compost container will rapidly fill up with cleaned annual weeds, grass clippings, leaves, and kitchen trash during the first growing season of your new vegetable garden. If you can, find it in or near your food garden. In this manner, you won't have to go far to transport materials for composting from the garden and finished compost back onto the soil.

5. Begin Your Planting Plan

Now that your growth regions are prepared, you may start your planting strategy. When establishing a new vegetable garden, my first thought would be to plant crops that would remain in the ground for a long time, such as bushes and fruit trees, as well as perennial vegetables like rhubarb and asparagus. Since you won't want to relocate them after they're planted, take your time deciding where to put them. The best place for permanent plantings is close to the garden's perimeter, keeping in mind the shade that larger plants, such fruit trees, would throw.

For the other vegetables, begin by ranking what you like eating first and, more importantly in smaller gardens, what will produce the most, such as pole beans, zucchini, salad leaves, and tomatoes..

Plant the right fast-growing plants.

These are a few of your favorite garden crops that mature quickly.

Lettuce. The majority of lettuces like the chilly early spring and autumn weather. Certain types may even withstand a little cold without any issues. For an even early harvest, you may start the seeds inside or immediately sow them in the ground. A wonderful veggie that grows quickly is lettuce. In only 45–50 days, you might be enjoying a fresh salad straight from your garden!

Arugula. Ready to harvest in around 50 days, arugula is a terrific addition to any salad. Although it's one of the costlier greens at the grocery store, arugula is simple to produce yourself. When the leaves are two to three inches long, harvest them. You'll have an ample supply of leaves throughout the season if you cut the leaves since this will encourage the plant to produce more.

Cucumbers. Planting cucumbers too early in the spring is not advised because they cannot withstand frost. If you want to start your cucumbers sooner, you may start them inside. Once the first set of true leaves develops, you can transfer them into the garden or container. You will be able to eat fresh cucumbers in fifty days. You may either let the cucumbers sprawl out on the ground or set up a trellis to make plucking them easier. There are types of cucumbers designed especially to grow in containers. Remember to plant some for harvesting later in the season. Keep an eye on that frost date!

Radish. After sowing the seeds, radishes may be ready to harvest in as little as 30 days, making them the crop in your garden that grows the quickest. They go well with fresh lettuce salads or on the veggie platter because of their spicy taste. Try roasting radishes in the oven with a small drizzle of oil and salt to taste for a novel take on radishes.

Pak Choi. This crop, which may be harvested in as little as 50 days after seed planting, also likes chilly spring and autumn temperatures. Use cooked or fresh pak choi.

Beets. Certain types of beets can withstand a mild frost, demonstrating their extreme cold tolerance. About 60 days is when beets are ready to be harvested, although you may pick young beets sooner if you'd like. Beet greens may also be picked and used to salads or soups. To aid in the development of the root, use the outer leaves and always leave four or five leaves. Beets may be cultivated all winter long in warmer climates.

Carrots. If you want to enjoy fresh carrots within 50 days after planting, choose for types that are just half grown. These cultivars are also suitable for container planting on a patio or balcony.

Kale. Growing kale is really simple, and when you pluck the leaves from the plant, more grow. There are many distinct types of kale, and each has a somewhat different taste and leaf form. Harvesting kale may occur as soon as 40 days after sowing. Consume kale raw, cooked in soups or casseroles, or combined with other salad greens. Not only are kale chips tasty, they're also rather healthy. Because it's so beautiful, kale may be grown as an edible decorative in the flower border.

Okra. Okra may be harvested 50 to 60 days after sowing, depending on the cultivar. Don't plant okra too early in the spring since it enjoys the summer heat but dislikes the winter. These plants are quite prolific, and if you keep picking when the pods begin to develop, you should be able to harvest continuously until the first frost.

Spinach. As soon as the soil can be worked, early spring is a good time to plant spinach, a crop that grows well in chilly climates. In less than 30 days, spinach may be harvested and added to mixed salad greens or served wilted with bacon dressing on its own. Spinach may winter over and begin to produce new leaves early in the spring, making for a very early harvest if you plant a late autumn crop.

Green Beans. After the risk of frost has passed, plant green beans, and in less than two months, you may harvest fresh beans. Fresh beans from the garden have the best taste and each plant may yield an incredible amount. If you don't want to trellis your beans or prefer to grow in pots, consider bush types.

Zucchini. After planting, zucchini may be harvested in as little as 40 days. Once there is no longer a chance of frost, plant the seeds straight in the ground. The zucchini will keep producing until the frost if you keep it harvested. But use caution! It will grow to the size of a baseball bat in a matter of days if you miss one. But do not worry! It provides a reason to make some delectable zucchini bread!

Mustard Greens. Another cool-season crop that may be harvested in a little 20 days is mustard greens! Plant the greens in the spring and again in the autumn, since the warm heat will render them bitter. Mustard greens may withstand a mild frost, although they are not as cold-tolerant as kale or collard greens.

As you can see, many veggies may be harvested in as little as two months, and in certain cases, even less time. As you garden, you'll discover what thrives in your area and how long it will take a certain plant to achieve maturity given the climate and grow zone. One benefit is that you may space out your planting so that no crop is available to harvest at the same time.

Soil preparation and planting

One of the most crucial elements in growing a good vegetable garden is preparing the soil. Fertile, well-drained, but wet, and receiving enough air circulation—all essential for strong roots—are the qualities of the ideal garden soil.

Explore more about the varieties of soil in your area.

The best time to start preparing your soil is in the early spring. Starting with a soil test will provide you with information on the pH (a measure of acidity and alkalinity) as well as the amount of nitrogen, phosphate, and potassium in your soil.

You may determine what soil amendments are required for the greatest plant development based on these data.

Steps in soil preparation:

The following are the measures to take while establishing a new garden site:

- Take away any grass.
- If you are establishing a new garden site from an already-existing grassy space, this is what you do.
- Rototill, shovel, or plow the ground.
- Make sure the soil is prepared for work; if it is, the soil structure will be harmed. Take a handful of dirt and squeeze it to assess the quality. It is too moist if it remains in a ball. It is too dry if it is powdered or has stiff clumps. It is perfect if it crumbles easily. Another indication that the soil is too moist to deal with is if, while turning the soil

with a spade, the dirt adheres to the end of the shovel. If the soil is very damp, wait a week before repeating the test.

Take down any sizable clusters.

After rototilling, there could be some remaining clumps. Use a hand cultivator or garden fork to break them up.

Apply soil amendments and manure or compost.

Using a hand cultivator or rake, add well-rotted manure or compost and tamp it into the top few inches of soil. Add 1/2 to 1 inch to the Roots area, 2 inches to the Brassicas region, and 2-3 inches to the Everything Else area. Visit vegetable garden plan to find out more about the aforementioned groupings.

Find out from your soil test whether any of the fundamental nutrients are lacking in your soil and make the necessary additions.

Add in lime

Just the area where your brassica group will be planted should be limed. In the event that your soil has a very low pH (below 5.5), you should still lime the "everything else" group but not the "root" group. The ideal time to lime your garden soil is one month after adding compost or manure. Apply 6 pounds for every 100 square feet of clayey soil, 4 pounds for loamy soil, and 2 pounds for sandy soil.

Rake the bed.

The purpose of this is to level and smooth the bed. Remove any big stones or rubbish. Ideally, the soil should have the texture of coarse breadcrumbs, particularly if you are planting veggies that have little seeds.

The next stage in growing a vegetable garden is to choose and plant your vegetable seeds when your soil preparation is finished.

Vegetable Seeds

- Choose reliable seed providers for your veggie seeds.
- Determine how many seeds you'll need depending on the design of your food garden. Next, decide which cultivars will flourish under your circumstances and submit your purchase. To ensure that you have the types you choose, you may do this early in the spring.

- After the earth has been prepared, begin planting your seeds.
- Time management is crucial while growing veggies. With this simple-to-follow book, you can remain organized by keeping a garden notebook.
- Seeds may be planted immediately in the garden or inside in trays or flats by the home gardener.

Starting vegetable seeds outdoors:

- When the weather is suitable for that specific veggie, sow the seeds. The seed will decay in too-wet or too-cold soil.
- Using a stick, hoe edge, or your fingers, draw the row to the depth suggested on the seed package. A seed planted too deeply will not germinate.
- If the soil is dry, water it.
- Along the row, scatter thinly. Never pour straight from the package; instead, transfer the vegetable seeds into your palm and hold them between your thumb and finger. You will be in more control of how much is planted in this method.
- Use your hand or the back of a rake to gently replace the dirt over the seed to cover it. Soil should be applied to the seed twice its size as a general rule of thumb.
- Using your hand or the bottom of a hoe, firmly compact the seed bed.
- Apply a little mist of water to the seed bed; an excessive amount of water might wash the seeds away. Don't let the soil become too wet. You may need to water your beds twice a day until the seeds sprout in really hot conditions.
- After two to four weeks of growth, thin the plants.

Starting vegetable seeds indoors:

Certain vegetable crops should be started inside, especially those that are hard to produce outside or need an extended growth season.

You have better control over the temperature, moisture content, and light levels when you grow inside. You also begin the growth season with a head start.

Most nurseries sell kits for beginning seeds. Sterilized potting soil and containers with drain holes, such as peat pots, cubes, pellets, or paper cups, are the essentials. Use a diluted mix of one part bleach to ten parts water to clean the used containers.

- Firm the soil to just below the top edge after adding the potting soil mix to the pots.

- Wet the dirt in the pots by adding water.
- Plant seeds at the depth suggested by the seed package.
- Fill the dirt around the seeds. As a general rule, the seed should be covered to a depth twice its size.
- If the temperature is higher than 20 degrees Celsius for seeds to germinate, use a heating pad or wire.
- As soon as the seedlings grow, give them plenty of sunshine.
- Don't let the containers become too wet.

Pot on: After four to six weeks of development, some plants, like tomatoes, must be potted into a large pot.

Putting the transplants into the garden bed:

It's crucial to move your seedlings from the inside to the outside when transplanting them. The procedure known as "hardening off" entails putting the plants outside on a warm day and progressively increasing their outside time until they are left alone. Usually, this procedure takes a few days to a week.

- Before you begin planting, make sure the pots are wet.
- Aim to avoid upsetting the roots.
- For the plant, dig a hole and add water to the bottom.
- After placing the plant in the hole and filling it in around it, use your fingers to gently press it down. Verify the proper depth for each plant, since some need a greater depth than others.
- Don't allow the plants dry up; keep them well-watered.
- Protect the plants, here are 8 ways to protect your vegetable transplants from extreme heat or cold.

Buying transplants:

Your local nursery will offer a wide selection if you decide against growing your own plants. Make sure the vegetable plants you choose are:

- compact and bushy
- leaves and stem are a healthy colour
- free of any insects or disease

- not spindly or leggy
- not limited by the roots; roots shouldn't protrude into drain holes

After establishing a vegetable garden, the following step is to keep the plants growing.

Vegetable protection

A crucial factor to take into account while designing your garden is vegetable safety. Depending on the crop and the growth circumstances, protection is sometimes required when placing out your transplants to keep them warm in chilly temps or sheltered from intense heat. Different measures of protection may keep your vegetables happy and healthy while also preventing pests and illness from wreaking havoc in your garden.

Vegetable protection ideas: Use this list to keep your plants safe from heat, cold, frost, and pests. Find out more about beginning transplants and seeds for vegetables.

1. cardboard container To enable it to be opened during the day, cut the bottom on three sides. On top of the plant, place the box upside down. shields against insects, birds, heat, and cold.
2. Tin cans Remove the can's two ends; a big coffee can works well. Plant it firmly in the surrounding soil. will shield from the cold and wind and keep out pests.
3. A wooden fragment To protect seedlings from the sun, place a board or shingle in place. If you put young transplants on a bright day, use this to shield them.
4. Paper goods Slice off the paper bag's bottom. Place four stakes around the plant and fasten the sack to them using staples. will shield your plants from the heat, wind, and cold.
5. Reusable plastic bottles Take a plastic milk jug and cut off the handle and the bottom. To anchor it to the ground, drive a stake through the handle hole. In warmer conditions, avoid using plastic containers. will shield seedlings from birds, snails, wind, and cold.
6. Cotton sheets Cover the plants with old bedding or drapes, then take them off throughout the day. Since they are lightweight, the plants won't be harmed. Use dirt or pebbles to tack down the cloth's corners. will shield plants from freezing temperatures.
7. Carpet collars or tarpaper Make 3-inch squares out of the tarpaper or carpet. In order to fit the square around the plant's base, trim it on one side now. This will lessen the likelihood that cutworms and cabbage root maggots will cling to the plant.
8. Remay:

Remay is a thin cloth that shields seedlings from insects, the elements, and the cold.

This will keep carrot rust flies, cabbage worm butterflies, thrips, flea beetles, and cabbage root maggot flies away from your vegetable transplants.

To create a little greenhouse, it may be placed directly over the plant or fastened to plastic tubing.

9. Wood or plastic pipe framework may be covered with or fastened to using plastic sheeting. It could become too warm, so be careful to open it throughout the day. used to shield plants from freezing temperatures.
10. Fencing will prevent animals like deer, raccoons, cats, dogs, and rats from entering your garden. Fencing may be made of wire, metal, electric, or wood.

Maintain plant growth

After you have planted your vegetable seeds and transplants, it is critical to continue the development of the plants. Taking the time to go around my garden and observe was the finest advise I ever had from a farmer.

Take some time each day or two to monitor your veggie plants. Keep an eye out for indications of damaging pests such as aphids or snails, as well as garden insects.

For most plants, a little amount of damage is acceptable; but, if you can detect it early, your plants will be healthier and more fruitful.

Your soil type and plant observation can help you determine whether additional conditions are necessary for healthy plant development. Watering, weeding, appropriate fertilizing, and plant protection are some of these needs.

Tips for great plant growth:

Watering:

To maintain consistent development of your vegetable plants, you will need to water them regularly enough. Insufficient water will make your plants wilt and eventually destroy the roots; enough water will do the same thing by drowning the roots.

Weeding:

All a weed is a plant that grows in an unwanted place. Many of the species that we consider weeds are really highly useful to our plants and environment. See my companion planting page for additional information.

Keep weeds away from your seedlings and budding seeds, even if certain weeds have advantages. Young, developing vegetable plants may be deprived of moisture and nutrients by them, which are essential for healthy development.

When the earth is somewhat moist and the weeds are little, it is ideal to remove them. Weeds should never be allowed to grow to seed. Mulch may be used to help protect vegetable plants after they are larger.

Weeding techniques:

The safest technique to pull weeds next to plants is by hand. When pulling weeds, always sure to remove the whole root.

It is better to hoe in places that are not near vegetable plants. Maintain sharp tool blades.

Take out and dispose of the garden's weeds in the compost. It is better to dispose of weeds that are in blossom or have gone to seed since their seeds could sprout in the compost.

For expansive garden grounds, using machinery (such as a tractor or rototiller) is beneficial. Weeds that are overturned may reappear; they are meant to be controlled rather than completely eradicated.

Fertilizing and Amending:

In order for vegetable plants to develop properly, they need nutrition. It might be necessary to use various soil fertilizers to give your plants an extra push throughout the growth season. Once the plants are well-established, which is typically two to four weeks following planting, do this. For information on when and how to fertilize your veggies, see my vegetable garden diary.

Some gardeners mix together soil amendments, which are used to enhance soil texture, with organic soil fertilizers, which are used to boost soil fertility.

Fertilizing techniques:

As a side dressing, either gently rake in the dry organic fertilizers with your hands (this is probably the simplest method) or scatter well-rotted compost around the base of the plant and water it in.

Pour 1/4 of a bucket or trash can full of compost or manure to make compost tea or manure tea. Pour water into the container, let it rest for at least a day, then dilute the liquid (1 part tea to 2 parts water) and use it to water the plants.

Protection:

There might be a range of weather throughout the growth season. It may be necessary to protect plants from wind, pests, bigger animals, extreme heat, cold, or frost.

Trellis and staking:

Certain plants, such snow peas, pole beans, tomatoes, and cucumbers, grow best when they are staked or trellised.

You may prevent fruits from contacting the ground and save space by using trellising or staking. When planting, provide supports to prevent subsequent root disturbances. As the plant develops, train and knot it.

Trellis and staking techniques

- wire cages.
- sticks or bamboo for tepees.
- robust wood frames.
- tall bamboo, metal, and wood sticks.
- polystyrene pea arbor.
- the thread between robust sticks.

Probably the most enjoyable part of growing a vegetable garden is the following stage! Find out more about the produce you grow.

Harvesting and preserving produce

The fruit of your labor and the time you have invested in your garden is a successful and bountiful crop of vegetables. The best part is undoubtedly getting to eat fresh vegetables.

It's critical to continuously monitor your plants to determine when they're ready to be harvested. The veggies may not have the desired size, flavor, or sweetness if picked too early; if picked too late, the vegetables may become rough or mushy and lose their flavor.

To have the finest flavor, veggies should be harvested when they are at their peak.

How do you know when to vegetable harvest?

Every vegetable plant produces food in a different method and has a distinct lifespan. While some grow food below, others create it above ground.

When veggies are grown above ground, you can observe how they appear, giving you an idea of when they're ready to be picked. Moreover, you can feel how full they are.

More challenging are those found below earth. For instance, potatoes may be picked at any time of year; they just fluctuate in size.

Other tips to for harvesting:

Some, like maize, you only ever harvest when they are at their ripest.

Others, like beans and Swiss chard, you can harvest several times if you pick them often.

Certain vegetables, like celery and carrots, remain in the garden longer than others.

For example, kale and cabbage taste better after a frost.

Certain plants—like tomatoes—preserve better than others.

Others, like spinach, don't keep well in storage.

Storing:

You could have more vegetables than you can utilize or share now that you have begun your harvest. How do you use the leftovers? Vegetables must be stored properly to maintain their nutrition, flavor, and freshness.

Keep these veggies cold and out of direct sunlight:

- sweet potatoes
- rutabaga
- winter squash
- onions
- potatoes
- Keep these veggies fresh in the fridge in a crisper or a closed plastic bag:
- asparagus
- beans
- cabbage

- carrots
- cauliflower
- celery
- beets
- broccoli
- corn (if husked)
- scallions
- turnip
- cucumbers
- leafy greens
- leeks
- radishes
- zucchini
- parsnips
- peas if shelled
- peppers
- In the refrigerator, uncovered:
- peas in pods
- corn in husk
- Store at room temperature:
- Tomatoes
- Preserving:
- Pickling, canning, and freezing are typical methods for preserving your produce.

Conclusion

Vegetable gardening is really fulfilling. To discover what will grow best for you in your individual garden area, have fun and be open to trying new things.

Enjoy your time in the garden, don't worry about mistakes, and use them as motivation to attempt something new or different the next season.

There are a few things to keep in mind if you want your garden to be generally successful. For a plant to grow successfully, it needs care and attention. It's crucial that you decide on a vegetable gardening strategy that suits you the best. If you don't have a garden, consider growing in pots on your front porch or balcony.

You may have a vegetable garden of whatever size you like or need. A good garden starts with carefully considering what you want to plant and how you want to cultivate it.

BONUS

Amazingly Abundant Crops To Grow In Your Garden!

BOOK 9

ORCHARD OASIS: NURTURING YOUR FAMILY'S FRUIT TREES

Fruit trees are a wonderful addition to any garden; with proper care, they will provide gorgeous blossoms and lush foliage, but they will also yield delicious, organic, home-grown fruit, making them highly useful. But if they aren't kept up, they may become quite dirty, grow scraggly, and become an eyesore. Fruit gardening doesn't have to be difficult. This guide will assist you in choosing, planting, and caring for your own fruit trees in your garden that are suitable for beginners.

Why plant trees for fruit?

Fruit gardening is an enjoyable, fulfilling, and, dare we say it, profitable endeavor. At the conclusion of your labors, you not only reap the rewards of a harvest, but you also maintain control and are aware of precisely what ingredients go into your food throughout manufacturing.

Decide On A Type Of Fruit

With so many types and possibilities available, selecting the ideal fruit tree for your garden may be daunting. However, you can reduce your options by taking into account your hardiness zone, soil type, sun/space needs, and the fruits you really love eating. Apples, pears, cherries, and plums are among the most popular fruit trees to plant if you live in a chilly area.

There are several cultivars (varieties) of each of these fruit kinds that you may choose from. Apples may be light yellow or deep red in color, and they are usually sweet or sour in taste. Apple types that are in vogue include Granny Smith and Honeycrisp. There are Asian and European types of pears. Asian pears are usually rounder in form and have a crispier texture, whilst European pears are longer and have a softer feel. Depending on whether they are sour or sweet, cherries have different characteristics. Sour cherries need to be collected sooner than sweet cherries since their skins take longer to soften. Additionally, plums come in a variety of unique types. Japanese plums, for example, are often tiny and have green skin, whereas hybrid plums are somewhat bigger and resemble plums from the grocery store.

Selected fruits that you like eating and can turn into jams, jellies, pies, and fruit leathers are just as crucial as the sum of the other considerations. You won't take the time to properly care for your tree and harvest the fruits when they are ripe if you don't love the fruit. When making a final decision, do take the time to thoroughly brainstorm with your family.

Whatever kind of fruit tree you choose for your garden, it's important to learn about each variety's traits before making a decision so that you can receive the greatest results for your environment. There are so many varieties to choose from, you're sure to discover the ideal one for your yard!

How to grow fruit trees

To begin with, determine the precise number of trees required to produce fruit. Many fruit trees have trouble producing blossoms of their own. In order to pollinate and bear fruit, many trees of various, suitable varieties that bloom at the same time are needed for apples, sweet cherries, pears, apricots, tangerines, mandarins, and plums. The exceptions include tart cherries, peaches, and most other citrus fruits, which have the ability to self-pollinate.

Supplies you will Need:

- Shovel/garden fork
- Tree support & ties
- Hammer
- fencing in the event that there are nearby deer or rabbits

Firstly, a warning: While planting may be done by one person, it is simpler if assistance is provided.

Bare-root trees: Before planting, soak the roots in a pail of water in a somewhat shaded area for one to two hours. Cut out any damaged or fractured roots.

Container-grown trees: Take out of the saucepan with caution. Splay out the roots as much as you can and shake out as much of the growth medium as you can.

1. Loosen the Sidewalls of the Hole

You may do this by using a spade or shovel to cut the side of the hole, or by inserting a garden fork into the sidewalls in the same way that you would tenderize meat. As the tree develops, loosening the sidewalls makes it easier for the roots to pierce the surrounding soil.

2. Remove any Labels

Remove them to stop the trunk from girdling as it gets bigger.

3. Place the Tree in the Hole at the Correct Depth

It matters how deep you plant your new fruit tree! Too much depth is often used when planting trees, which seriously impairs healthy development. Apple plants are grafted onto a unique rootstock, which is essentially a meticulous union of two distinct trees—the fruiting and rooting portions—to produce a hybrid tree.

The top fruiting portion of the tree will produce roots via a process known as scion rooting if this graft is positioned below the soil line. This will negate the grafting process and leave the tree with a weaker root system.

Grab a friend, and let's start planting:

- Position the root ball in the planting hole so that the visible point of the graft union, where the fruit tree and rootstock were grafted, is just above the soil's surface. To make sure of this, laying a piece of wood, a bamboo cane, or a fence post over the top of the hole will allow you to see the ground level after the hole has been filled in.
- While you start filling in behind the roots so the tree may rest in the bottom of the hole without assistance, have a fellow hold the tree at the proper height in the hole. Spread the roots evenly and start filling up the remaining portion of the hole after the tree is able to rest in it.

4. Fill in the Hole

When you fill the hole, gently press down with your foot to eliminate any remaining air pockets, being careful not to crush the dirt around the roots. As you get to the top of the hole, make a little depression or bowl to let water naturally collect around the tree. Take care not to pile dirt around the trunk.

5. Provide Support

Until the trunk of a freshly planted tree becomes strong enough to hold itself, it has to be staked during the first two to four years after planting, particularly if the tree is placed in an area that is prone to severe winds.

Drive the tree support, either perpendicular to the ground or slanted slightly away from the tree stem, into the hole's edge with your hammer, being careful to avoid as many roots as possible. For the first three feet of the tree's trunk, fasten the tree to the support. Use rubber or old nylon stockings or any other soft material for your ties.

6. Water the just planted tree.

Despite soaking your bare-root tree before planting, all trees need watering right away. Water it with two to five liters. Move carefully; add extra water after letting it soak. By removing any air pockets that could have developed during the filling-in process, the water helps settle the soil around the roots.

After watering, you may need to add additional dirt or compact it once again.

7. Mulch Around the Base of the Tree

By minimizing weeds and preserving soil moisture in the root zone, mulching may lessen competition that might hinder the development of young trees. Apply completed compost or well-rotted manure to a depth of about 1 inch, extending outward to a distance of around 3 feet from the trunk's base.

Mulch should never come into contact with the tree's bark and should always be kept at least 3 inches away from the trunk of the tree.

8. Keep rabbits and deer away from your young tree.

Your tree runs the danger of being nibbled on by Peter Cottontail and deer until its bark hardens. Your tree may get damaged or perhaps die if they eat the bark. The best method to keep animals away is using fence, including chicken wire and other kinds.

A lower fence near the trunk will suffice if the only visitors to the tree are rabbits; deer, on the other hand, need taller fencing separated a few feet from the tree to prevent them from reaching the upper branches as the tree develops.

9. Prune your Newly Planted Fruit Tree

Pruning promotes side-branching the next spring and root development. Pruning a tree to a height of thirty to thirty-six inches is a decent general rule of thumb. Many apple and pear trees are trimmed to develop a single, central leader as they mature. Timing is another important factor in pruning success.

With time, the top bud will develop into the main stem or central leader of the tree. Take off any branches that extend below the tree's lowest 18 inches.

Pro Tip: Lacking room? Discover how to properly prune your fruit tree. Espalier is a practical way to save space when used up against a wall or fence.

10. Give it plenty of water the next day.

The next day, give it another 2–5 liters of well-watered water. Basic tree maintenance then takes over, while trimming apple trees calls for extra attention. If you live in a location where there isn't a weekly rainfall of at least one inch, then start a weekly watering program for the tree.

Where To Plant Fruit Trees

The optimal location for planting fruit trees will rely on the particular variety's needs for shade and light. The majority of fruit trees, however, like full sun. Fruit trees need to be in direct sunlight for at least eight hours per day.

Fruit trees do not do well in very alkaline soils; instead, they do best in sunny, sheltered spots with well-drained neutral to slightly acidic soil.

You should also think about how the placement of your fruit tree may effect garden upkeep. Choose a planting spot close to your home's landscaping where the trees may flourish and produce fruit without creating a mess.

Fruit trees are best planted in the winter, when they are dormant.

Container-grown plants may be planted at any time of year, but winter is the ideal time to plant them. Bare root trees should be planted between November and early April.

Planting trees is most economical in the winter. Fruit trees planted bare-root in the winter cost less than half as much as those grown in pots in the spring.

One of the greatest plants for a tiny garden is a lemon tree in a terracotta pot.

Planting Fruit Trees In Pots

Dwarf fruit trees are great potted tree options and may be used to create small garden ideas by setting them on a patio or balcony.

Size matters a lot when it comes to fruit trees grown in containers.

Get the largest container within your means, then plant your bare-root tree in the winter with plenty of garden soil, compost, and earthworms added.

Fruit trees need to be anchored when placed in pots in order to be protected from strong winds. Most fruit trees like full sun and shade, but check with your particular tree to be sure.

Fruit trees that are grown in pots need continuous spring and summer fertilization and watering since dry plants will shed their flowers in an attempt to live. Check them daily and feed them once a week with a specialized fertilizer for fruit trees.

How Do You Prepare The Soil For Planting Fruit Trees?

Understand how to test the pH of the soil before preparing the ground for the planting of fruit trees. You may do this by using a home test kit or by sending a sample of soil to your nearby institution to be examined in a lab setting.

Fruit trees should ideally match the pH of your soil naturally, however you may adjust the pH of your soil somewhat using lime or acidifier. For instance, plum trees like 6.0–8.8 pH, whereas apple trees want 5.5–6.5 pH. Therefore, if you get the pH correct, your efforts will pay off in every season.

Select a dwarf variety that you can plant in a container if your soil type entirely conflicts with the requirements of the tree, even if little pH changes are achievable.

In order to guarantee that your tree receives the nutrients it needs to flourish, it's also critical to gradually enrich the soil with compost. Instead than simply adding potting soil or compost to the hole, amend the whole soil area with it. As a result, tree roots will be encouraged to spread out into the surrounding soil rather than curl up in the hole.

How Far Apart Do Fruit Trees Need To Be Planted?

The distance at which fruit trees should be planted is somewhat determined by whether they need to be cross-pollinated or are being taught to climb a trellis.

To prevent root competition and to let light reach the ground, fruit trees shouldn't be placed too near to one another.

They should generally be spaced between 10 and 30 feet. Permit a minimum of 6 feet to separate espaliers and 2 feet between cordons.

What Fruit Trees Should Be Planted Next To Each Other?

Whether or not cross-pollinating is required for your selected tree will determine which fruit trees are best planted adjacent to each other. In such case, you'll need a minimum of two types.

In order for cross-pollination to produce fruit, self-fertile trees don't need another nearby, but pollinating trees require the presence of at least one partner tree.

For instance, a crab apple tree will pollinate most apple kinds while bearing apples.

Fruit Tree Care

General Care

Three things should be taken into account while selecting a location for your home orchard: sunlight, soil, and spacing.

SUNLIGHT

John Denver was content in the sun. It will also provide happiness to your fruit tree. Choose a spot for your tree that gets at least half a day's worth of sunlight. The fruit produced by the tree is more abundant when it receives sunlight. Avoid planting your tree in a completely shaded spot.

SOIL

Fertile, well-drained soils are preferred by fruit plants. Most soils drain enough to keep your trees healthy. However, before planting, add one-third more peat to the soil if it contains a lot of clay. This will assist in improving your tree's drainage. It is necessary to stay away from full clay soils and poorly drained areas. Low, moist areas of your yard with inadequate drainage are not conducive to fruit tree growth. You may plant your tree or trees on top of a mound or berm made of trucked-in dirt if your soil is exceptionally heavy and poorly drained.

SPACING

We have chosen all of our trees to be dwarf or semi-dwarf in order to maximise available area and provide bountiful fruit yields. Planting trees should be done 12 to 14 feet apart. If planting many rows, space them out by 18 to 20 feet. This will give the tree plenty of room to grow and flourish. The light may beam down on the tree from this place. Additionally, it offers enough air ventilation, which lessens the risk of infections on your tree.

Finally, while choosing the location of your orchard, don't forget to take your future ambitions into account. Make space for more trees. As soon as you begin cultivating fruit at home, you should add more varieties to your crop.

POLLINATION

Blossoms are pollinated to produce fruit. Self-pollinating refers to a kind of tree where the pollen from the tree itself may produce an abundant harvest. Some trees need pollen from a different species. Bees are often responsible for this cross-pollination. In some neighbourhoods, there are sufficient fruit trees to provide abundant cross-pollination; nonetheless, it is advisable to establish additional "pollination partners" for further assurance. Two trees of the same variety will not cross-pollinate if the variety is not self-pollinating.

A pollinator is usually needed for the majority of apples, pears, plums, and sweet cherries, while some of those fruit kinds have self-pollinating variations. Nearly often, peaches, nectarines, tart cherries, and apricots self-pollinate.

For comprehensive information on the best pollinators for your apples, pears, plums, and sweet cherries, look under the respective fruit categories. Recall that pears cannot fertilise plums, and apples cannot pollinate pears. While cats and dogs do not breed, pollinators need to come from the same fruit species.

PRUNING

It is essential to emphasise the significance of consistent, yearly, and forceful trimming. Preserving the tree's continuous vitality and optimising fruit yield are crucial.

The tree's final form is determined by trimming in the first year. After planting, cut your tree to a height no more than 4–6 feet above the ground. Remove any branches that are crossing over one another and any inwardly developing branches. To promote growth, trim off the tops of the bigger branches. View the figure below for a comparison of the branches before and after.

Prune any branches or shoots that emerge from BELOW the "bud union" both now and in the future. Suckers are young stems that emerge from the ground and root systems. Just chop them off at ground level if you spot them. Generally, suckering decreases as the tree ages.

Should your trees bear fruit in their first year, remove a few of the immature fruits from the branches, placing them about 8 inches apart. This will promote healthy ripening, provide good spray coverage, and enhance vegetative vigour. In the future, fruit thinning will be crucial for the same reasons. Extra is less. If you don't thin, the tree won't be able to withstand the large number of fruits you acquire, which will lead to damaged branches and undersized fruits. So don't be scared to becoming skinny. The fruits that are produced will be considerably better and fuller.

It is beneficial to "shape" your tree in later years. The finest trees to train to a central leader are apple, pear, and cherry trees (uppermost upright limb). It is best to train peach, nectarine, plum, and apricot trees into a vase form without using a central leader.

WHEN TO PRUNE

APPLES AND PEARS

Pruning apples and pears is usually best done in their dormant state. So choose a bright, beautiful winter day and enjoy this aspect of orcharding. Pruning in the summer helps to slow down a tree's development. Therefore, prune the tree in July to stop its aggressive growth if it is becoming too tall for you.

CHERRIES

Cherry tree pruning is usually best done in the summertime. Winter, late autumn and early spring are not the best times to prune. All non-arid settings have bacterial infections, which are especially harmful to sweet cherries. The chilly, rainy season is when these bacteria are most active. Therefore, do not trim your cherry trees until the tree has flowered and the warm, late spring weather patterns have been entrenched, which is often by the end of May.

PEACHES, NECTARINES AND APRICOTS

Early spring is the ideal time to prune apricots, nectarines, and peaches. Consider trimming after your region's final date of frost. By trimming off the majority of the winter damage now, you may lessen the impact of late frost damage on your buds and flowers.

PLUMS

Pruning plums should be done vigorously since they develop extremely quickly. Remember that trimming your plum tree in the summer, when it is still developing, can help control its tendency to spread. A plum tree can never be pruned too much. Thus, tidy up your trimming in the winter to remove any dead or damaged branches and give your tree a more symmetrical form. Then, to keep the size moderate, prune once more in July.

Maximizing fruit production

Imagine having a beautiful view of your home from stunning trees that also greatly benefit your family – fruit!

Fruit trees are among the most satisfying plants to cultivate and own since they are aesthetically pleasing and have a high nutritional content.

Here are some suggestions for improving the production of your fruit tree, whether you're just starting out or have been cultivating one for some time and are not happy with it. Read on down below!

1. Fertilize Regularly

Do you like your trees to provide an abundant crop of fruit? Fertilizer is essential to doing this.

They will get the nutrients they need from this to provide fruit that is healthy. By doing this, you may ensure that they don't lose their fruit crop and end up with an abundance of fruit.

2. Plant Your Fruit Tree in an Area With Full Sun Exposure

After you've decided what kind of edible garden would produce the most for you, it's critical to consider where on your property to put it.

However, you can think about hiring a permaculture garden designer to really assist you ensure that you can use your available area. They will create a plan that is specifically tailored to your requirements and the location of your home, helping you to improve the produce from your edible garden.

3. Prune Annually to Encourage Fruit Production

Dead and decaying branches are cut off during pruning to make way for new growth. This is another excellent precaution you can take to prevent harm to your property and to onlookers.

Pruning preserves the plant's natural structure and encourages healthy development by warding off insect and animal infestation. Pruning trees and shrubs promotes a healthy crop of fruits and flowers.

4. Water Regularly

The movement of nutrients and carbohydrates from the soil to the trees depends on water as well. Every plant, including trees, will eventually experience a drought or lack of water.

Remember to bring water with you since this may sometimes be deadly or drastically hinder the development of those plants.

5. Choose a Suitable Variety for Your Climate

It's important to choose fruit trees that are appropriate for your environment.

Choose a fruit that you like and that will grow well in your growth zone before you visit the fruit tree nursery.

6. Be Vigilant for Pests and Disease

All gardeners will probably come across pests and plant diseases at some time. Your plants may not always be in danger from these things. When you can, get rid of pests, and as soon as you can, carefully remove any sick plant portions.

Depending on what, how much, and where they are developing, each person has varied demands. However, you must take into account your requirements for a fruit tree agricultural sprayer.

7. Cross-Pollinate

Compared to self-pollination, cross-pollination aids in increasing fruit output.

Because bees are excellent for this, choose plants that attract bees to assist take care of the bees as well. If you are willing, you may even begin tending to a nest yourself.

Be Patient With Your Fruit Trees

In gardening, more than in many other areas of life, patience is a virtue.

Fruit tree varieties develop, blossom, and bear fruit in their own lovely pace. It's time to start thinking forward to the coming season and envisioning what it would be like to eat local fruit.

You may increase the produce from your fruit trees by paying attention to these pointers. Check out our most recent pieces on this subject if you like this one.

Harvesting

Possess fruit trees laden with abundant fruit? These pointers will assist you in harvesting the rewards of your hard work.

When you shake the limb, the fruit comes off. It is fastest and simplest to use the first method. A big sheet under the fruit tree and a few eager folks are all you need. Select a branch that is loaded with fruit, then shake it well. Naturally, everything falls from the tree as it is shaken, including the leaves and other detritus, so you will need to separate the excellent fruit from the bad.

Use a basket for picking fruit. Use an extended handle on your fruit-picker basket for easily pulled-off, hard-to-reach fruits. You may pull on the fruit by encircling the hooks around its stem. Right into the basket goes the fruit.

Chop off fruit. When it becomes difficult to remove some fruits by tugging, use portable pruners. Use telescoping pole pruners for those sensitive fruits that are difficult to reach. Pole pruners have the extra benefit of holding while they cut, which makes them ideal for delicate and tender fruit like ripe peaches.

Pull and twist. Pomegranates, for example, are not as easy to remove off the tree as apples are. If you are without pruners, take each fruit by itself and twist it till it comes off the tree.

After harvesting has finished, examine the fruit and remove any poor ones to compost. If you have excess fruit that you would want to contribute, find out the timings and locations of any local harvest services by calling your local food bank or doing some internet research.

 Why not endeavor to preserve the produce or jam-making if you have a couple of hours to spare?

Recipe

How to make fruit jam

When you have extra fruit from your fruit trees, try these delectable jam recipes; they'll make wonderful drink presents for loved ones:

Ingredients:

- 2 kg apples, cleaned
- 3 cups water

- 1 lemon
- 1,5kg white sugar
- 1 tablespoon cinnamon

Method:

1. Scrub, wash, and chop the apples before adding lemon juice to the water.
2. Cook the apples in water with 0.5 kg of sugar until they become mushy.
3. Cook the boiling apple mixture for a further ten minutes after adding the remaining sugar.
4. Add the apples and stir in the cinnamon; turn off the heat.
5. Transfer heated jam into glass jars that have been cleaned and sterilized, seal them, and place them in a boiling water bath for ten minutes.

Homemade plum jam

Ingredients:

- 1 kilogram of plums
- 600g of sugar
- One teaspoon of vanilla
- extract from one lemon

Method:

1. Boil the plums for two minutes, then remove and cover with cold water to facilitate easy peeling.
2. After peeling and halving the plums, remove the pits and place the halves in the enamel pot. After covering the plums with sugar and lemon juice, let them cool overnight.
3. Cook your prepped plums for around twenty-five minutes at a moderate temperature the next day.
4. Stir in the vanilla when the fruit jam has simmered. Fill heated, sterilized jam jars with hot jam, then screw the lids on firmly.

FAQs

Q: I have a hectic schedule and not much time for gardening. Which kind of fruit tree is your recommendation?

A: A great option for you would be dwarf fruit plants. Because they are compact in size, they are simpler to prune and harvest. They also take up less room, which makes them ideal for patio pots or little gardens.

Q: How much space do I need to grow a fruit tree?

A: The kind of tree you choose will determine how much room is needed. Larger trees, such as a full-sized apple or pear tree, will need more space to develop than dwarf trees, which may flourish in a smaller space or a big container. Espalier trees may grow flat against a wall or fence, making them an excellent choice if you want to maximize vertical space.

Q: I live in a colder part of Australia. What type of fruit tree would be best for my garden?

A: Certain fruit plants, such as apple and pear trees, do well in colder climes. Think about going with a cultivar that has a reputation for being cold-hardy.

Q: How much maintenance do fruit trees require?

A: The kind of tree might affect the degree of maintenance. Every fruit tree need periodic fertilization, consistent watering, and some degree of trimming. Additionally, certain fruit tree species could need extra attention, such as insect management or particular pH values for the soil.

Q: Can I grow different types of fruit trees together?

A: Of course! In order to create a varied and abundant garden, many gardeners plant a variety of fruit trees together. Just make sure the trees you choose demand the same amount of sunshine and irrigation.

What is the estimated duration needed for my fruit tree to yield fruit?

A: Everything relies on the kind of tree you plant and how mature it is at that point. While bigger trees may take three to five years to begin bearing fruit, some dwarf trees may begin doing so in as little as one to two years.

Recall that the pleasure of gardening comes from selecting options that align with your own interests and lifestyle. You can locate the ideal fruit tree for your yard if you take the time to look around.

Conclusion

In conclusion, there are seven crucial actions you should do to guarantee success if you want to cultivate fruit trees in your garden. Finding your USDA hardiness zone is the first step in order to better understand the kinds of fruit trees that grow in your location. Next, determine if the soil type in your garden or yard is appropriate for the sort of tree you want to plant. Based on its size, estimate how much room and light exposure the tree will need. You should also consider the kind of fruit you would want to eat and look into certain varieties that thrive in your area. Ultimately, it's critical to buy from a respectable nursery.

Having all the information you need before purchasing the tree will assist guarantee that it grows healthily and produces high-quality fruit for many years to come. growth fruit trees successfully also requires proper care; be sure to water them frequently but not excessively, trim them when necessary, fertilize throughout the growth season using organic techniques wherever feasible, and apply natural oils that repel pests as needed to keep them away.

Remembering these pointers can help ensure that you have an abundance of fruits growing in your own backyard, including delicious pears, luscious apples, and even rare species like quince! Plan ahead before you plant! Homegrown fruits provide unmatched flavor that cannot be topped, whether they are used as components in delectable dishes or to fulfill those mid-morning snack desires! Thus, don't hesitate any longer and begin choosing your ideal piece of nature right now!

BONUS

Growing Fruit Trees in ANY Size Garden

BOOK 10

HARVESTING THE WILD: A GUIDE TO HUNTING AND FISHING

During the last two million years, hunting has been essential to human existence. First of all Hunting is a fun activity and one of the few methods to get fresh organic meat in the modern world. Unlike when you purchase at a food store, you do not have to worry about antibiotics, hormones, or steroids when you harvest an animal while hunting. Hunting not only gives people access to a good diet but is also a crucial technique for controlling animal populations, since many species have no natural predators.

What is hunting? How to start hunting? How do I get into fishing? How to prepare game?

To provide answers to these queries and aid in the success of hunters and anglers, this book was written. There are sections below that cover a wide variety of hunting activities. Every section will have information on safe hunting techniques as well as helpful hints on how to hunt that specific species. There will also be advice on processing game that has been collected after certain portions.

Hunting

Hunting is a time-honored activity that may be done for trophy collection as well as food gathering.

There's something about hunting's essentially primal qualities that gives folks a confidence and pride boost. That being said, it ought to be done with respect and without harming the environment.

Let's go over a few hunting techniques in this book guide to get you started with this age-old pastime.

Interesting Facts About Hunting America

In the United States, hunting is more than merely bringing down large antlers for a hunter's wall or obtaining sustenance. Any novice hunter should know these fascinating hunting facts, since there are many to learn. These are a few of the most interesting hunting facts.

Comparing it to store-bought beef, it is healthier. Meat purchased from stores is usually obtained from animals housed in cramped quarters and given artificial diets to gain weight. Selling an enormous, meaty animal is, after all, more lucrative than selling a healthy one. Any animal you kill in the wild will have lived its whole life on natural nutrition, making it much healthier than any store-bought meal.

It is really beneficial for conservation. Proper regulation of hunting may actually contribute to the survival of endangered animals. However, it has to be strictly monitored.

Not every animal is hunted for food or awards. Both people and agricultural animals are at risk from some bigger predators. Hunting such animals is the only method to deal with the issue when it gets out of control.

Although there are many other hunting methods, the majority of American hunters use a rifle or a bow, and if they want to lead a primitive lifestyle, they may also employ traps. In general, blowguns and other instruments that are used in other nations are not utilized in the United States.

You may use a helicopter to hunt. The US government permits hunters to kill animals from helicopters when they pose a threat, albeit this practice isn't as popular as the other ones. Boar hunting is among the instances of this. When boars are causing damage to a farmer's property or a natural resource, the authorities have the authority to let hunters to fly above the area and take immediate aim at the animals.

How to Start Hunting

Hunting is a difficult but worthwhile activity. But, you wonder, how can I begin hunting?

Those of us who had the good fortune to have a hunting parent as a parent had the added benefit of being able to use the lessons and observations we picked up early in life.

It might be difficult for those who weren't raised in a hunting household to know where to begin. You should use this step-by-step hunting beginner's tutorial.

1. Ask yourself why you want to hunt

If you're going to hunt, be sure your motivations are sound. Aiming to be a nasty person? Not pleasant.

Do you want to wow your partner or give yourself a little more machismo? Not good at all.

Are you hoping to get some organic, lean meat for your family and discover more about wildlife? There's a strong incentive to hunt now.

Please do yourself a favor and research the advantages of hunting, then determine why these advantages are significant to you.

2. Take a hunter's education course

The best place to start is with a hunter's education course, which will cover safety, ethics, techniques, and more. By completing an online hunter education course, you may get your certification as a hunter. Nonetheless, if it's feasible for you, I suggest taking an in-person course, since it will provide you the chance to interact with your teachers and classmates and handle weapons firsthand. To expand your hunting knowledge after finishing your foundational course, think about spending money on a graduate-style school like a master hunter or hunter advancement course.

3. Do your research

Before you head for the hills, give yourself the benefit of a little knowledge and get on the books. All of the subjects listed here are worth reading:

Species Of Game, Tactics For Hunting, Ethics Of Hunting, Archery, Firearms, Ammunition, And Wildlife Habitat, Hunting Birds, Suiting Up For A Big Game, History Of Hunting, Cooking

Wild Game. Most of this subject listed here is been discussed down below but do well to read more on each subject.

4. Find a mentor

Although it's not a must to start hunting, having a mentor might be beneficial if you're a novice. If you don't personally know any hunters, how do you go about finding a mentor?

Attending an in-person hunter's education session would probably help you meet individuals who would be pleased to mentor you. If taking an online course is your only choice, consider volunteering for a conservation organization, going to a local conservation banquet.

5. Study the regs

I know—reading the hunting laws manual for your state sounds about as enjoyable as eating dirt. However, you should definitely take the time to study them carefully, paying particular attention to the hunting districts and species you'll be hunting, unless you want to be a fool and make foolish errors that might cost you your hunting license, money, or even time in prison.

6. Choose your weapon

Initially, choose whether you want to hunt small game, large game, birds, or any combination of these. Choose the kind of hunting you want to perform as well. Will you use hounds, a bird dog, a blind, or a tree stand? Are you going to be spot-and-stalk large game or shooting ducks from a boat? Are you going to hunt outside of your state? Select your weapon(s) according to your budget, style, and intended target.

Rifle: more effective than archery at longer ranges and for hunting large animals. Typically, rifle season occurs later in the year and is cooler than archery.

Shotgun: useful for game birds like ducks, grouse, and turkeys. Coyotes and deer may be taken with shotgun slugs.

Muzzleloaders: are more archaic weapons that may be used in places where guns are prohibited since they can only fire at closer ranges.

Crossbow: Depending on your state, you may use it in the firearms or archery season. Out of bounds in Oregon.

Conventional bows: rely on instinctive aim. Two varieties: longbow and recurve

Compound bows: are more widely used and often have more power than conventional bows.

7. Practice

Whichever weapon you choose, be sure to invest a lot of time on it for the animal's sake as well as your own safety. Although accidents happen to the best of us, no ethical hunter wishes to injure an animal. Repeatedly practice, repeat, repeat. Make use of various targets, field settings, and dry-firing tactics. Prior to the commencement of hunting season, have your bow or handgun sighted in.

8. Plan your pack-out

For bird hunters, it's not a major concern, but if you're hunting larger game, you really need a pack-out strategy. Motorized vehicles are not always permitted for the collection of game. To pack away their meat, many hunters use horses, mules, llamas, or goats. You will have to act as the pack mule if you are animalless. If the terrain is cart-friendly, then game carts are useful. Buff buddies are also beneficial!

9. Get boots on the ground

Not only is it beneficial to break in your boots, but there are other advantages as well.

You must become in better condition. Even if it's possible, don't count on shooting an object from 200 yards away from the vehicle. Aim to wander a lot and dispose of the meat by either packing out what you shoot or feeding it to pack animals.

It is essential that you acquaint yourself with the hunting landscape. Is it mostly level or steep? Did it fire in the previous years? Is a supply of water present? If you were in good form, would it be difficult to pack out an animal because of the amount of cover or is it that thick?

It's important that you watch the animals, particularly the ones that you want to shoot. Keep an eye out for predators and other moving objects while surveying the area. Examine the signs. Are there fresh scat, tracks, beds, wallows, rubs, or game trails? Is there a suitable location for a tree stand or blind, if any?

10. Gear up to start hunting

It's not necessary to own the priciest or most elegant equipment; many old-timers still hunt in jeans and plaid shirts. But having the right gear—at the at least, suitable footwear and clothes for the conditions and terrain, a good hunting pack, a knife, game bags, bear spray, a comfortable weapon, and an emergency kit—will make hunting more enjoyable. Binoculars and a rangefinder are two more items that some hunters would consider necessary hunting equipment.

Should I dress in camouflage gear?

Depending on what you're trying to find. Turkeys and ducks may be valuable since they can blend in with their surroundings and have exceptional eyesight. Deer, elk, and the majority of other game bird species, however, are colorblind, so they may not see your vivid red coat.

11. Plan to be safe

Show at least one person where you're going using a map or GPS location, and let them know when they may expect you back.

12. Be your own butcher (or pay someone else)

Let's assume that after all of your preparation, practice, research, and real hunting, you manage to catch an animal. With great care, you field dress it, removing as much flesh as possible, and get it home before it goes bad. What happens next? The animal has to be butchered, so you can either do it yourself or take it to a nearby butcher to be processed.

13. Enjoy the meat and memories

Nothing is more fulfilling than settling down to eat succulent, wholesome meat that you yourself sought out and collected. If it doesn't taste well, you could be cooking it incorrectly.

How to Fish

The now is the ideal moment to begin your fishing journey. Learning new skills and spending time outside with friends and family may be achieved via fishing. You may quickly start reeling in fish and creating new experiences with only a few pieces of basic equipment.

Step 1. Get a fishing license.

You must receive a fishing license in the state where you want to fish before you can begin fishing. If you buy your fishing license online, you may even go fishing the same day.

Step 2. Learn to identify the fish species that inhabit your state waters.

It's essential to learn how to identify the species you capture. Once you have mastered the ability of correctly identifying fish, you will be able to consult any size restrictions, bag limitations, or other rules about fishing for that specific species with accuracy.

Step 3: Find a nice fishing location.

To choose your site, use the best family locations list or the places to boat and fish map. It might be useful to confirm access points, prospective structure, or fish activity if the place is close to you and you have time to visit it before your real fishing excursion.

Step 4. Assemble your fising gear and tackle.

Remember that all you need to start fishing is a little amount of tackle and equipment. Make use of this basic list of beginner-friendly fishing essentials:

- combination of rod and reel.
- Tiny tackle box with trays that are split.
- Monofilament fishing line spool (for freshwater use, use an 8- or 6-pound test).
- Round plastic bobbers.
- Divided shot weight.
- a variety of hooks in sizes ranging from size 3/0 for larger bait to size 2 for smaller bait.
- Pliers (for taking out hooks).
- Measurement tape.
- Scissors (for cutting a leader or line).

Step 5. Pick up and master a few basic fishing knots.

Learn one knot for linking lines and another for rigging or lures. The improved clinch knot (for tying your leader to your fishing hook) and the double-uni knot (for combining lines) are two useful knots to start with.

Step 6. When going fishing, make sure you are aware of the state's fishing restrictions and pack a copy.

It is essential that you be aware of any applicable fishing rules or limitations for the species you harvest. The purpose of fishing laws is to safeguard our rivers and fish populations for the enjoyment of future generations.

Step 7: Obtain some living lure.

You may get your own live bait or buy it from a bait or tackle store. You may use live worms, crickets, or minnows as effective freshwater fishing baits.

Step 8. Head to your fishing spot and bait your hook.

You know precisely where you want to go for your first fishing adventure since you have previously done your homework on local nice fishing sites. When you get there, unpack your equipment and use the live bait you packed to bait your hook. Remember to use the hook that is appropriate for the kind of bait you are using.

Step 9: Throw your line into the ocean.

To get your bait at the right spot in the water column while using a plastic bobber, you may need to modify where it is placed on your fishing line based on the water's depth.

Step 10. Wait for a bite.

When using circular hooks for fishing, keep in mind that you only need to reel in order to set the hook—you don't need to jolt your rod upward. You must swiftly raise your fishing pole into the air to set the hook if you are using a normal hook. How wonderful that you now have your first fish on the hook!

You are now familiar with how to go fishing, so make sure you also read the following subsection on how to release the fish you catch. Since you won't want to keep every fish you catch, it's essential to understand how to release fish to maximize their chances of survival.

HOW TO RELEASE A FISH: STEPS

"What's the best way to handle fish to ensure they survive release?" is probably a question on your mind. The angler has a major influence on whether or not they do. It takes practice to learn how to release fish from a hook in general.

- Handle fish with watery hands or knotless rubberized landing nets (if you have to use gloves, wear rubberized ones, not cotton ones). This keeps the fish's slime coat in good condition, which keeps it safe from illness and facilitates swimming. When handling fish, anglers that understand correct catch and release techniques never use any form of towel since it might remove the slime covering.

- Since fish travel through the water in this manner naturally, hold the fish horizontally whenever you can. Avoid letting the fish fall onto rough areas!Avoid putting your fingers near the fish's eyes or gills.

- To reduce handling, employ a release tool (dehookers, recompression tools) if necessary.

- There's not much time left! Don't keep fish out of the water for longer than absolutely necessary; release them as soon as it's practicable. In order to assist the fish breathe again, try releasing it gently head first into the water. This will force water over its gills and into its mouth. Fish that are tired may be revived by putting them in the water, preferably facing the stream, and holding their tail or lower lip with one hand while supporting their belly with the other.

To ensure that you follow the rules, be aware of the most recent fishing restrictions that are in effect in the state in which you are fishing, and acquire the skill of measuring fish precisely. You are sustaining and increasing the number of fish in our state for fishing enthusiasts to come by adhering to fishing restrictions.

Game and fish preparation

The meat might be ruined by mishandling wild animals in the field or preparing it poorly at home.

Fall marks the beginning of several wild game seasons for hunting.

The utilization of the prey they capture is something that hunters eagerly anticipate. Meat may be ruined, however, if it is handled carelessly in the field or cooked poorly at home.

The way meat is handled throughout the process, from harvest to preparation, may significantly impact the final product's safety and taste, according to a food and nutrition expert.

Here are some pointers for managing big game animals in the wild, including deer:

- To guarantee the quick loss of body heat, dress out (remove the entrails) as soon as it is killed.
- Using a dry or wet towel, clean the cavity that was gutted. Attempt to maintain the animal's cleanliness.
- Using a stick to prop open the chest cavity and let air flow freely, quickly and completely cool the corpse. Additionally, hanging the carcass helps it cool.
- Purchasing ice cube bags to insert inside the animal's body cavity may help cool it down on a hot day. Make sure the ice is still inside the bags.

After shooting game birds, as soon as you can, remove the entrails and crop them. This facilitates rapid and complete cooling of the corpse by allowing air to flow throughout the body cavity. Put each bird in a plastic bag and set it on ice if the weather is too hot. Avoid grouping warm birds together.

Fish also require some careful handling:

- As much as you can, keep the fish you capture alive. A live box or a metal link basket are preferable than a stringer. Avoid tossing fish into the boat's bottom.
- Put them in an ice chest filled with ice if you are unable to keep them alive.
- As soon as you can, clean them.
- The wind is going to dry out the fish, thus keep them covered while you fish in the winter.

Fish and game meats might help to diversify your diet. **Some tips may be used when preparing game meat:**

- You may use bison, venison, elk, and beef interchangeably in recipes.
- Game meat is often drier since it has less fat. To make up for it, try marinating it before cooking it like a steak or stir-frying it, baking it in oven bags, or using it in soups and stews.
- Safely handle wild game. For optimal quality, keep raw wild game frozen for up to six months or refrigerated at temperatures below 40 degrees Fahrenheit for up to two days.
- Place frozen meat in its original packaging and thaw it on the lowest shelf of the fridge. Place the meat in a waterproof wrapper in cold water to expedite the thawing process. Replace the water as required to maintain the chilly temperature.

Hunting and Fishing Safety Tips

We like being outside, thus we like to hunt, fish, and shoot archery. Although engaging in any of these three activities may be thrilling and enjoyable, doing so carries some risk.

According to an examination of hunting incidents in 2018, there were 17 incidents in total, three of which resulted in fatalities. Even while the number of instances may appear low, the unfortunate truth is that many of them might have been avoided.

This list is a set of safety guidelines for each of the three activities to assist in preventing mishaps and injuries.

Hunting Safety Tips

This hunting season, if you are planning a vacation, be careful to tell someone about your plans. Tell them when you expect to return so that, in the event that you don't, they will know to seek for you.

Keep in mind that your jacket, shirt, and cap must all be blazing orange. You will be noticeable to other hunters because of this hue.

Gun Safety

1. Ensure that you know how to use your rifle. Make sure you understand how to load and unload it as well as how much kick it has when the trigger is pulled.
2. When hunting, you should always presume that all of the firearms are loaded. As a result, aim the muzzle in a secure direction.
3. Never aim a gun toward or in the direction of a person or animal.
4. Until you are ready to fire at your target, leave the safety on.
5. Put your finger on the trigger only when you are ready to shoot.
6. Make sure you know exactly where you want to aim and that there are no people immediately in front, behind, or to the side of it.

Tree Stand Safety

1. It always takes two people to install and maintain a tree stand. Always work in pairs!
2. Examine the straps and the stages. Replace any component that seems worn out or broken.
3. When in a tree stand, wear a harness at all times.
4. To raise your bow or guns, use a rope. When you draw up the rifle or bow, be sure they are not loaded.
5. Always carry a mobile phone so you may contact for assistance in the event of an accident.

Fishing Safety Tips

Perhaps the finest outdoor activity is fishing. Fishing excursions are not only enjoyable, but they also include camping, boating, outdoor living, and a wonderful opportunity to connect with nature.

Your first concern should always be safety while engaging in outside activities.

Boating Safety

1. Ensure that the boat is equipped with all the necessary supplies for your fishing outing and a first aid kit.
2. Acquire proficiency in using the boat's rescue equipment, such as life jackets, flotation devices, and flares.
3. Be careful not to overburden the boat and to equally distribute the weight.
4. Restricted locations and low-water dams should be avoided at all costs.
5. Before you go, check the weather. If you are aware of an impending storm, stay off the water.
6. When on a boat, wear a life jacket at all times.
7. Make sure your boat lights are on if you are going at night so that others may see you.

Tackle Safety

1. Examine your surrounds before casting to prevent your hook from snagging on a tree, power wire, or worse, a person!
2. Your tackle should never be left on the ground.
3. When the hook is deep within the fish, use a hook remover or cut the line back as far as possible. To remove the hook from the fish, avoid reaching within its mount with your hand.
4. Never leave hooks or lures on the line; always keep them in your tackle box.

Archery Safety Tips

Bows and arrows are deadly weapons, and you should treat them with the same care as you would a rifle, whether you use them for hunting or as a pastime.

1. Crossbow and arrow should always be pointed in a safe direction.
2. A bow should never be nocked until it is safe to fire.
3. Don't ever aim over a crest.
4. Make sure you know what's in front, behind, and to the side of your objective.
5. Don't shoot unless you have a secure backdrop or backstop, as well as a safe shooting range.
6. An arrow should never be shot straight up into the air.
7. When using a bow and arrow, always use finger protection and an armguard.
8. To prevent injury, store arrows in a covered arrow quiver.

9. Always look for damage to your bow, such as dents, cracks, breakage, or any kind of mechanical malfunction. If a bow seems to be damaged, never shoot it.
10. Any arrow with irreversible defects should be discarded.

Conclusion

There's no denying that hunting is a closely watched and often misinterpreted activity. Regarding the ethicality of hunting, there are many, and very strong, views. Those who oppose hunting usually make the first point that there is never a necessity to kill an animal.

However, a closer look reveals that's not the case. Many people who do not hunt, and even some serious hunters, are unaware of the many, many advantages of hunting. Hunting has considerably more benefits for the environment, ecology, and economy than just the activity itself. In actuality, hunting is essential to the preservation of species and the habitats in which they reside.

To help spread the word about the advantages of hunting, here are a few justifications for its positive effects.

Conservation

The fact that hunting generates revenue for wildlife conservation is one of the main, if not the main, justification. The sales of hunting permits and associated fees generate millions of dollars annually, all of which are used to support the preservation of wildlife.

To keep wildlife populations healthy, wildlife conservation works to preserve wild creatures and their natural habitats. Funds for conservation also support the preservation, improvement, and restoration of natural ecosystems.

Moreover, an excess tax was established by the 1937 Pittman-Robertson Act and is levied on all sales of weapons, ammunition, and archery supplies. More than $1.6 billion in funds is raised annually for wildlife conservation in the US thanks to the Act.

Population Control

Another crucial aspect of maintaining healthy animal populations is hunting. The more people we live in cities, the more of a negative impact our human imprint has on animals as cities expand.

Unfortunately, the natural balance between predators and prey is upset when humans intrude on nature. Without any natural predators to control their population growth, this leads to a sharp rise in certain animal populations. For instance, an excessive number of deer indicates a reduction in food availability. This may result in ill and undernourished deer.

The overabundance of predators is comparable. Herbivorous animals like deer would have fewer in number if there are too many predators. As a result, the ecosystem's natural equilibrium may become disturbed. For all species to coexist, equilibrium must be maintained.

Economic Value

In every prosperous sector, including hunting and fishing, money is always a major component. The yearly economic impact of consumers spending their hard-earned money on outdoor leisure in the United States is $887 billion. An annual tax revenue of almost $125 billion is generated by this. Furthermore, 7.6 million Americans are employed in the sector itself.

The figures remain astounding even when we expressly divide the two businesses apart. Consumers who go hunting and fishing spend about $63.1 billion at retail each year, and almost 483,000 Americans work in these fields.

Hunting Combats Poaching

Poaching is the illegal hunting and/or trapping of animals on unclaimed territory. It also includes species that are completely prohibited from hunting or that are killed outside of their designated hunting seasons.

The first line of defense for reporting poachers and the waste of wild wildlife is the hunter. Because game wardens are unable to monitor every square inch of territory, hunters are essential in the battle against poaching.

Since there is a purpose behind the hunting and fishing rules, true hunters respect and follow them. This reduces the immoral and illegal slaughter of animals by those engaged in poaching. Seasons and regulations pertaining to hunting are designed to safeguard sportsmen as well as animals.

BONUS

Hunting & Gathering

Book 11

Fortify Your Home: A Comprehensive Defense System

Entry into your house might be detected by home security systems. A siren sounds and, if you have a professional monitoring service, a professional monitor calls to confirm that the alarm is genuine after a short pause to let you deactivate the system.

Let's examine popular choices for home protection, including security sensors, monitoring services, protocols for communication, and more.

What is a home security system?

In order to defend against thieves and other possible house invaders, home security systems are networks of integrated electronic devices that collaborate with a central control panel.

A basic home security system includes:

- A base station or control panel with an independent keypad.
- sensors on windows and doors.
- sensors for motion.
- breaks in glass sensors.
- surveillance cameras.

However, there are a ton of other gadgets out there that may improve the security of your house, such as environmental sensors and smart home goods.

How Do Security Systems Work?

All of the parts of most home security systems are connected to a base station. They then establish a connection using a smartphone app, which gives us remote access to monitor and manage everything. Most systems link to our phones via Bluetooth, Wi-Fi, or cellular service. In the event of a power loss, some systems choose to include a landline or battery backup.

We arm our security system as we leave. We disarm it when we get home by using our smartphone, a voice command, a key fob, or entering our password on a keypad. Our camera immediately stops recording and none of the sensors activate when our system is disabled. On the other hand, regardless of whether our security system is on or not, our smoke and carbon monoxide detectors stay on. Naturally, our cameras begin to record and the sensors activate once our system is armed.

We can chat using two-way audio, see live video from our security camera, get alerts from our sensors, and do other functions using a mobile application. Unlike a neighborhood alarm system, smart home security enables us to monitor our house from anywhere.

How Does Alarm Monitoring Work?

What would happen then if our armed alarm system was set off by someone? Our base station receives a message from our sensors instantaneously, and relays it to our mobile application. We get email notifications, SMS notifications, or notifications inside the app, depending on the system, the app, and the settings on our phone. After that, we either broadcast live video of the events or converse with the person in front of the camera using the two-way audio feature on our camera. We notify the police whenever we become aware of an intrusion.

Naturally, due to our hectic schedules, we are unable to constantly respond to our security alerts. For this reason, a lot of businesses provide round-the-clock, expert system monitoring. If one of our alarms goes off and we are unable to handle it, the monitoring center's expert monitoring staff is alerted. They check the situation and then make our emergency services call.

It is advisable to have expert monitoring available around-the-clock, as nobody can always be there to oversee their home's security. However, self-monitoring is also an effective choice if we want to avoid monthly or annual expenses. These days, a lot of systems also allow for DIY installation. Our expenses are considerably reduced by self-monitoring and do-it-yourself systems. With these kinds of systems, the equipment is often the only expense.

For those who wish to self-monitor, it is much advisable to purchase a siren that will activate in tandem with the alerts. This siren may be installed on the base station or be a separate device. The siren need to be at least 85 dB, or around the same volume as a diesel truck. Recall that in an emergency, it is best to make your noise louder so that your neighbors can hear it too.

Home security measures

Installing security systems is undoubtedly less enjoyable than furnishing your new house. However, because there is a burglary every 30 seconds, home security needs to come first. These easy steps will help you secure your new home quickly so you can get back to the exciting stuff.

Here are 11 easy ways to secure your home

- Set up a security system
- Secure the doors
- Lock the windows
- Light up the landscape
- Don't forget the garage
- Lock down your Wi-Fi network
- Eliminate hiding places
- Add security cameras
- Get a safe
- Use home automation
- Prevent house fires

1. Set up a security system

Whether it's a do-it-yourself system or one with expert monitoring and home automation capabilities, your new house needs protection. For every budget and degree of protection, there are many different home security alternatives available today.

You consider the requirements of your home and community before deciding on a system that you are comfortable with. Examine your house to determine what needs to be done to safeguard it when you are aware of the possible threats.

For information on area crime rates and assistance in conducting a home security assessment, get in touch with your local police department.

Additionally, bear the following points in mind:

Expert installation vs do-it-yourself installation.

smart home features.

Monthly and upfront expenses.

Customer support and the standing of the brand.

extras like carbon monoxide and smoke monitoring.

2. Secure the doors

Stay away from assisting a thief who enters via the front door (34% of them do!). Check that the door frames on all of your external doors are sturdy, that the hinges are secure, and that, in the event that your door has a mail hole, someone cannot reach through it to open the door.

Change the door locks if you're moving into a place that someone else formerly called home. In this manner, you can ensure that your locks are the finest available and prevent outsiders from having a key to your home.

We suggest the following basic fixes to help you fortify these crucial entrances.

- Put in a deadbolt.
- Include a striking plate.
- Invest in smart locks.
- Increase safety by using a video doorbell.
- Keep the mail slot secure.

Strengthen doors with sliding glass: Make sure your sliding doors are secured since burglars like them. To prevent the door from being pushed open, you may use a window bar or dowel in the track, if that's more your style. Add a glass break sensor or door sensor for a high-tech solution. These should deter criminals and notify you if the glass door is tampered with.

3. Lock the windows

An expert on theft, claims that windows are "a common entry point for criminals" and that a prior homeowner may have left them closed but unlocked. Furthermore, windows made by manufacturers often have fragile and ineffective locks.

Increase security with aftermarket window locks or key-operated levers if you're not fond of the way your window latches appear. However, you don't have to end there.

To further assist make your windows burglar-proof, these are some excellent suggestions.

» Use window security film to strengthen glass.

» Install sensors for glass breakage or windows.

» Put window bars in.

» Under first-floor windows, plant thorny shrubs (but keep them clipped).

4. Light up the landscape

Criminals such as thieves and vandals dislike being in the limelight. Make sure your outside illumination is enough to keep them away. Install lighting around the garage and other exterior buildings, along walks, and all around your front and back yards. Not only will you scare off would-be burglars, but you'll also lessen the possibility that you'll trip up the front stairs.

These pointers can help your outdoor security lighting become even more efficient:

» Use lights that detect motion.

» Utilize solar-powered lighting to save electricity.

» Set the timer for your outdoor lights using a smart outlet.

» Use smart light bulbs to program routines.

5. Don't forget the garage

Criminals are starting to use this entrance point to your house more often. Furthermore, it's likely that you have a ton of valuables kept in the garage, even if they are unable to enter your home. Make it your practice to lock all of the garage's doors, inside and out.

You may also think about storing your garage door opener inside your home. This will prevent a thief from taking it from your vehicle. Additionally, if the garage is secured with a security code, be sure to keep it a secret and never unlock it in front of neighbors, delivery personnel, or other individuals.

Here are a few more simple strategies to keep the garage safe:

- Get a smarter garage door opener as an upgrade.
- To conceal the goods inside, cover the windows.
- Use additional locks to secure your garage doors.
- Never again leave the garage door open by using home automation.
- Get a driveway alarm installed.

6. Lock down your Wi-Fi network

Your financial and personal information may be accessed over your home WiFi network. Additionally, using home automation increases the risk of a break-in at your residence. Criminals may have direct access to your house if your security system or smart home devices are linked to your Wi-Fi network.

However, you don't have to expose yourself. Employ these strategies to prevent hackers from accessing your home network.

- Protect your wifi network.
- Turn on WPA2 or WPA (Wi-Fi Protected Access) encryption.
- Your home network may be hidden and renamed.
- Employ a firewall.
- Install malware and antivirus software.
- Make secure passwords.
- One of the best methods to safeguard your internet connection is to use a virtual private network, or VPN.

Use parental control software to safeguard children: If you have little children, it's likely that they use the internet for both leisure and schoolwork. Use a parental control app, router, or software package to protect kids against internet predators and cyberbullies. You may also use these technologies to ensure device-free family meals and set screen time limits.

7. Eliminate hiding places

Although they may improve the exterior appearance of your home, trees and bushes can provide easy hiding places for robbers. Prune any nearby trees and plants that may be utilized as shelter. Instead, go for smaller shrubs and flowers. If there are trees next to windows, you should either remove them or provide more protection to those windows.

Additionally, give your home's outside some attention. Make use of these recommended practices to maintain security.

- Ladders and stools should always be stored.
- Sheds, gates, and other outbuildings should all be locked.
- Don't leave expensive items out in the open to attract robbers.
- Invest in security decals and signage even if you don't have an alarm system.

Remember to install warning signs at every entrance to your home, as research suggests that they might be just as useful as the alarm itself.

8. Add security cameras

Most likely, you've heard news stories about security camera video foiling burglaries and porch pirates. One home security system that serves as a deterrence and a way to get justice is this one. Security cameras may be used alone or as a component of a full-featured home security system.

Regardless of your choice, we advise using a security camera that connects to a smartphone app so you can see live video and save it in case you ever need to call the authorities. Be proactive in preventing hackers from accessing your webcams.

Check out our other must-have features:

- Local or cloud storage
- Motion detection
- Wi-Fi capability
- Two-way talk
- Night vision
- A weatherproof camera housing for outdoor use

9. Get a safe

Make sure your valuables are secure in case someone manages to circumvent your other home security measures. A safe kept at home may store everything from jewels to important papers like passports. A safe that can withstand fire, be waterproof, and weigh enough to prevent a robber from taking it with them is what you need.

To maximize the safety provided by your safe, adhere to following recommendations:

- Seek for safes that have redundant locks, meaning the same safe has two locks.
- Choose the appropriate safe size based on the assets you want to keep secure.
- Choose between an anchored safe and a portable safe.

Keep weapons secure in a gun safe: A burglar will often target weapons as desired things. Every year, over 230,000 firearms are taken during home invasions. A gun safe shields your family from the possibility of a fatal accident and keeps your guns out of the hands of criminals..

10. Use home automation

If you've been tempted to turn your regular house into a smart home, security is one compelling reason to follow through. Home automation gives you remote (or scheduled) control of lights, door locks, security cameras, smoke alarms, and other safety devices. You can get real-time alerts about suspicious activity so you can respond quickly and thwart potential thieves.

Here are some of our favorite ways to use home automation to increase security:

Schedule lights (and your TV) to turn on and off when you're on vacation.

Scare away porch pirates with two-way talk through a smart doorbell.

Get an instant video feed whenever someone walks up your driveway.

Check on a smoke or CO alarm and cancel false alarms from your smartphone.

11. Prevent house fires

Do what you can to prevent a fire in your new home.

- Verify that the CO and fire alarms that are currently in place are operational.
- When needed, install new detectors.
- Verify the fire extinguishers' charge and expiry date.
- If needed, get brand-new fire extinguishers.
- A professional chimney inspector should be contacted prior to utilizing the fireplace. Here are some additional safety advice for fireplaces.

Remember to create a fresh fire escape plan that takes into account the layout of your house and includes fire escape ladders for bedrooms on the second level.

Self-defense strategies

You can never be too cautious these days about who you let into your house or who you approach too closely while you're out shopping at the mall or in another public place. Kidnappings, kidnappings, and other threats might occur in a second if you let your guard down. These kinds of incidents, which unfortunately have been reported in our local community, may happen to anyone.

To lessen the likelihood of an assault, go over these self-defense suggestions for both inside and outside. Above all, be alert at all times.

Protecting Yourself at Home

1. Lock Your Doors

It's a good idea to keep your doors closed even while you're at home, particularly if you live in a remote area where it would be difficult to reach aid in the event that a hazardous burglar did show up at your door. If at all possible, check who's outside before opening your door using peepholes or windows. If you are unsure about the person at the doorstep and are unable to inspect it without opening the door, just give it a little gap.

2. Close Blinds and Curtains

Both the interior and the exterior of your house should be well-lit, but at night, outsiders shouldn't be able to see inside. Invest in shades or drapes that will block out light so that no one can follow your movements or determine whether you're not at home or asleep.

3. Enroll in a Self-Defense Course

Learning self-defense techniques and reviewing how to repel an assailant may help anybody. In most cases, these programs also include how to avoid being attacked with a knife, pistol, or other weapon.

4. Get a Dog

According to surveys, owning a dog at home helps discourage break-ins and burglaries. Consider adopting a pet from one of the many local animal shelters if you don't already have a furry pal.

Protecting Yourself Away from Home

1. Always Be Aware of Your Surroundings

Even while doing something as obviously safe and secure as going grocery shopping or getting a cup of coffee, you should never let your guard down. It is very uncommon for kidnappings and abductions to occur in busy public areas, therefore you should constantly be alert and aware of your surroundings. If you believe someone is following you, stay out of situations where they may easily assault you, notify security, or ask for assistance from anyone around.

2. Always Fight Back

Refuse to give up on your assailant until you can escape to a well-lit and frequented place. Seize every chance to harm your assailant in order to escape. Target the face, eyes, crotch, and knees, among other places. If you are unable to strike, scratch, or punch with your hands, push your assailant away with your hips before kicking with your feet or knees.

Self-defense teachers may show you a variety of techniques you can use to break free from an anyone attempting to confine you. But never give up attempting to hurt your assailant and flee to a safer area, even if you forget what you're "supposed" to do.

3. Carry Mace and Know How to Use It

Carrying pepper spray, often known as mace, is a smart idea as a constant safety precaution. If you choose to carry it, however, you need make sure you feel comfortable utilizing it. To determine how far it will spray and how near you must be to your attacker for it to be effective, practice spraying it at home.

Here are some pointers to remember while using pepper spray:

- To prevent spray from blowing back into your face, always spray downwind.
- Target the assailant's face. Apply two or three sprays and then take off.
- Refrain from giving your assailant the pepper spray so they may use it against you.
- When you feel as if you are in a dangerous place, always have pepper spray on hand. It won't help if you have to rummage through your handbag to locate it.

4. Trust Your Gut

Follow your gut and leave the situation if you get a terrible feeling about someone you're speaking with, or even if it seems like someone is following you too closely. Don't stress about coming off as ungrateful or cruel. What matters most is your safety.

Emergency response plans

Your first responsibility in a crisis is to protect your family. A significant incident will probably cause uncertainty, turmoil, and a scarcity of necessary goods. Businesses could shut, and first responders will be overworked. A little preparation may go a long way toward providing you peace of mind when it comes to disaster preparedness at home. Creating a Family Disaster Plan

The protection of you and your loved ones is the most important thing during a tragedy. Making a plan can assist to minimize the financial effect and guarantee that you can concentrate on what matters most.

Since no two disaster plans are alike, it's critical to take into account your particular situation, including your home's layout, the ages and mobility of each member, your communication routes, etc. Here is a basic checklist to get you started:

Know your surroundings & create a supply kit.

After a tragedy, gather everything you could need. Generally speaking, it is advised to pack enough supplies to last at least three days and up to two weeks. So that you can "grab and go," keep everything in transportable containers like backpacks or duffel bags. You may wish to carry a smaller one in your vehicle as well. Go to Ready.gov/ for a list of recommended things.

Know what emergencies are most likely to happen in your area.

This covers both natural disasters like hurricanes, tornadoes, and floods as well as man-made ones like emissions from power plants and chemical plants. The risks change based on whether you reside in an urban area or a more rural one. You may find out what the biggest hazards are in your region by contacting your local emergency management organization or American Red Cross.

Know the dangers in your home.

Examine your house and its surroundings for any hazards, such as leaning trees, outdated appliances, blocked exits, etc., at least once a year. By being aware of these dangers, you can keep an eye out for any indicators that anything unsafe is happening in or near your house.

Know your utilities.

If you're a renter, get to know how to quickly contact your landlord and utility emergency numbers in addition to knowing how to switch on and off your services, such as gas, electricity, and water. Once a month, check your smoke alarms, carbon monoxide detectors, and fire extinguishers. If the batteries are changeable, replace them annually or as directed by the manufacturer.

Identify your shelter plan.

Where should you go if there is severe weather to be safe? The location of your safe spot may vary based on the circumstances. Take a flood, tornado, and fire, for instance.

Make an evacuation plan.

Does everyone in the home know where to go in the event of an evacuation? At least twice a year, conduct emergency evacuation and fire exercises. Consider the possibility of packing a single carload or bringing just one luggage. Which would you choose up? Prior to anything happening in your home, come to an agreement on this and make adjustments.

Put together an emergency communication plan.

Who will be the person to whom essential information should be communicated? Who would take care of your kids in case you couldn't be reached? How are you going to convey that information?

Assign responsibilities.

Who is in charge of the family pet, for instance? What about papers that are important? If someone given a task isn't home, how will your emergency plan change?

Create a plan for special needs.

It's crucial to have a plan in place if you or anybody living with you has specific requirements, such as a handicap. Provide details in your supply pack about any requirements, prescriptions, backup equipment, etc. You may tell local emergency services precisely what kind of assistance you would require in an emergency by registering with them, as many of them do.

Post emergency phone numbers in an open and obvious place.

Ensure that everyone is aware of how to contact friends and family in an emergency. To reduce the number of calls you need to make, think about designating a specific contact who can assist in alerting the others on your list.

Know your evacuation routes and designate a meet-up location.

Where would your family assemble if a crisis happens while you're away from home and the phones are out of commission? Never disregard emergency response teams' warnings. When a road or structure is declared dangerous, never try to enter it.

Prepare for stress.

Stress is a given both during and after a calamity. It cannot be avoided, but the more ready you are, the more capable you will be to deal with it. Putting coping mechanisms into everyday practice might be an additional step toward bettering your stress management.

Home Security System Pros and Cons

After learning the basics of what a home security system is and how it works, let's examine the main benefits and drawbacks of making the investment.

Of course, the primary function of a home security system is to alert you to the presence of unauthorized individuals inside your house, but allow me to elaborate.

Pros of a Home Security System

Quick police contact: We may be certain that while we're not at home, the police or other emergency responders are contacted thanks to expert surveillance.

A system's existence alone may discourage burglars: The mere presence of a placard and window stickers alerting potential burglars to our security system may deter them!

Simple safety alerts: A home security system might be helpful if we often have problems like unintentionally leaving the front door open or forgetting to lock the rear window. With smartphone apps, entry sensor data may remotely indicate how safe our property is. In addition, notifications let us know when movement or a person is identified, so we can check in and make sure everything is well at home while we're gone.

Potential reduction in homeowner's insurance: We may put more money in our wallets since many home insurance companies give discounts if we purchase a security system.

Livestreaming: It's not just for criminal activity! It's also a terrific way to see how things are doing at home, such as our teenage niece's "small get together" or our cats getting into mischief while we're away. No matter where we are, we can always be at home thanks to livestreaming!

Cons of a Home Security System

Monthly payments: We often have to pay monthly fees if we want to add cellular backup or 24/7 expert monitoring to our system. These might cost as little as $10 or as much as $60.

False warnings Every now and then, we get some false alarms that have even caused the police to become involved. They are a waste of time and humiliating for everyone.

Possibly forget to switch on: This is more of a personal issue, but some individuals report that they struggle to remember to arm their home security systems before they leave, rendering the purpose of the system meaningless. Although we don't experience this problem, we can see how it may make a system seem unnecessary.

Conclusion

People become desperate amid disasters. When the legal system breaks down, you may need to defend your family and your possessions. Fortify doors, windows, and vulnerable walls in your home, set up booby traps to deter intruders, install security cameras, fortify your perimeter, surround your property with prickly vegetation, add a four-legged guard dog to your family, and construct a safe room as a last resort to strengthen your home's defenses. These defenses ought to be sufficient to maintain your house's reputation as a safe haven, but it's a good idea to be ready for anything and to have an escape route organized.

To sum up, this book has served as a guide for protecting our loved ones and properties. Practical measures like locking doors and windows, learning self-defense techniques, and talking about emergency preparations have all been addressed. The book also examined the benefits and drawbacks of home security systems, highlighting the need of moderation.

Recall that genuine security is a continuous process that adjusts to fresh difficulties. Applying the knowledge you've gained from this article can help you make your house secure and resilient. Decide which tactics are most crucial to you, then arrange for the other techniques to be included later. The best method to protect your house and loved ones is to be aware of possible security concerns and act quickly to address them.

BONUS

HOME HARDENING. Fortifying a Place of Residence

SCAN ME

CONCLUSION

By the time you get to the last pages of the "No Grid Projects Bible," you've already left the confines of traditional living behind. This book has been designed to provide you with the abilities, insight, and motivation required to successfully negotiate the intricacies of our dynamic world.

You have learned the art of wilderness survival, adopted sustainable lifestyles, and become an expert in the nuances of cattle husbandry throughout these pages. You've turned into a health defender, equipped with vital first aid knowledge, and you've used solar energy to cut off from the conventional electrical grid. You have developed abundance in your own backyard and refined your skills of self-reliance as a modern homesteader.

You've become more in tune with nature's primordial rhythms by going on adventures into the wild, recognizing edible plants there, and mastering the art of hunting and fishing. You've tended vegetable gardens and orchards, establishing a haven of locally grown, fresh produce. Your homestead and the safety of your loved ones are guaranteed by the extensive defense system you have constructed.

The "No Grid Projects Bible" is a call to action as much as a compilation of guidelines. It's a call to adopt a resilient, sustainable, and independent way of living. Remind yourself as you put this book down that you are still on a journey to live off the grid. Keep learning, growing, and adapting. By imparting your knowledge, you can create a community of people who share your values and are dedicated to leading purposeful, harmonious lives with the environment.

I hope this compilation will be a source of light for you as you move toward living a life that is less reliant on outside forces. The lessons in these pages apply to everyone, whether you're a seasoned homesteader, an aspiring adventurer, or someone looking for a more intentional and sustainable lifestyle.

We appreciate your participation in this life-changing adventure. May all of your efforts bear fruit, your crops be bountiful, and your bond with the land and one another strong.

I hope your life is full of resiliency, independence, and the satisfaction that comes with being off the grid.